Human Dimensions of Civil Engineering

Written to complement civil engineers' technical knowledge, this book explains the sociocultural contextual knowledge that civil engineers need if they are to be effective in their professions. Civil engineers design and build the world in which we all live. The decisions that they make can guide us toward a more sustainable society since the infrastructure that they create has a direct impact on how sustainably we are able to live. Sustainability is value-laden, however, and embedded within larger contexts. Whilst engineers are well versed in technical matters and the evaluation of physical contexts, their education often leaves out essential knowledge about the larger social, cultural, economic, historical, and political contexts in which they operate.

This book helps readers to understand contextual knowledge and why context matters – which is useful to engineering students and professionals who have found this topic absent from their education, who would like to understand contextual issues, and who would like to know why they should care. The book lays out essential sociocultural contextual knowledge for today's civil engineers, relevant across a wide variety of workplaces.

Kathryn Terzano is Lecturer in Sustainable Design at the University of Bristol, UK.

Human Dimensions of Civil Engineering

Context and Decision-Making for a Sustainable Future

Kathryn Terzano

CRC Press
Taylor & Francis Group
Boca Raton London New York

CRC Press is an imprint of the
Taylor & Francis Group, an **informa** business

First edition published 2024
by CRC Press
4 Park Square, Milton Park, Abingdon, Oxon, OX14 4RN

and by CRC Press
2385 NW Executive Center Drive, Suite 320, Boca Raton FL 33431

© 2024 Kathryn Terzano

CRC Press is an imprint of Informa UK Limited

British Library Cataloguing-in-Publication Data
A catalogue record for this book is available from the British Library

ISBN: 978-1-032-49070-0 (hbk)
ISBN: 978-1-032-49114-1 (pbk)
ISBN: 978-1-003-39219-4 (ebk)

DOI: 10.1201/9781003392194

Typeset in Times New Roman
by Apex CoVantage, LLC

Contents

Why Contextual Knowledge Matters

1

1.1 INTRODUCTION: BEYOND TECHNICAL EXPERTISE

Civil engineers design, construct, and maintain the built environment that is necessary for daily life to exist. Without civil engineers, the world would be very different and considerably less developed. The infrastructure that underpins modern society, including roads, bridges, sewer systems, water systems, electrical grids, airports, canals, and railway stations, is in large part thanks to civil engineers. Many of the everyday conveniences that we take for granted today would not exist without civil engineers. We would not have dependable transport options that get us from place to place safely and efficiently. We would not have large-scale renewable energy sources like wind, solar, and hydroelectric power, nor would many of us have access to safe and clean drinking water. In addition, many of the largest and most remarkable structures in the world, from soaring skyscrapers to huge stadiums to intricate long-distance bridges, exist because of the contributions of civil engineers. Civil engineering allows the world to be more advanced, efficient, and safe, advancing the built environment to meet the needs of modern society.

Civil engineers are tasked with acquiring the technical expertise and problem-solving skills to ensure that all of the aforementioned impressive feats are accomplishable and that the structures that they create are safe for society. To be accepted into a civil engineering course of study, university admissions generally require evidence of proficiency in mathematics and physical sciences such as physics and chemistry. These qualifications have evolved from

DOI: 10.1201/9781003392194-1

the time before formal academic programmes in civil engineering existed, when instead learning the trade mostly involved completing an apprenticeship to an experienced engineer or as a draughtsman in an engineering office. In such a setting, an aspiring engineer would gain the technical expertise required to plan and develop infrastructure projects, such as buildings, roads, bridges, and canals, and this experience may have been supplemented by training in architecture or surveying or technical education in mathematics or mechanics. Although technical aptitude and acquisition of technical knowledge may have been sufficient in the distant past when civil engineering as a field and profession was newly established, this is no longer the case.

Today's engineers still need to be experts in technical matters, but they also need to understand for whom they are creating the built environment. While a journalist writes for a specific audience – the readers of a particular newspaper or watchers of a specific news programme – the civil engineer creates for society at large. Unlike an identified audience for a journalist, who could be as specific as university-educated, middle-aged women living in a certain metropolitan area, the civil engineer's 'audience' is everyone who could reasonably be affected by the structure being created, maintained, or deconstructed. The civil engineer must ask themselves, for example, whether the replacement of a bridge with a new, larger bridge will induce traffic in the surrounding neighbourhoods and, if so, how that will affect the neighbourhood residents.

Without such understanding, civil engineers will fail to take into account the complexity of the modern environment. A holistic view of the context in which engineering works are created, including for whom these works are created, is necessary in order to make sound decisions about what needs to be created, when, how, and where. In engaging with their public audience, civil engineers should seek to understand the public's history, desires, and beliefs, and how the demographic makeup of the public is changing. This is not only to meet the highest purpose of serving the public, but also to make the job of being a civil engineer more efficient and effective.

Contextual knowledge is essential for civil engineers to understand the specific opportunities and challenges for a given project. To develop and implement buildings and infrastructure that meets local legal and regulatory standards, civil engineers need to be knowledgeable about the relevant laws, building codes, and zoning rules. To develop long-lasting infrastructure, they need to account for local climate and environmental conditions that might vary greatly from one place to another. Economic and social factors such as population growth, demographic changes, and economic development can affect the demand for infrastructure and the viability of different solutions, and to meet the needs of the local population, civil engineers need to be aware of these issues. Designing infrastructure can incorporate local culture and values, especially in areas where there is significant culture and a need for historic preservation, and civil engineers need to be mindful of this and should seek to collaborate with local stakeholders. Understanding the local environment can also enhance project outcomes, resulting in projects that cost less and are more appealing to the community.

Related to this is the concept of sustainability. The civil engineer has the responsibility and the privilege of making decisions that directly influence how sustainably people are able to live, whether the infrastructure is in place to take public transit or whether one will need to rely on roads and automobiles, the extent to which the quality of life is improved, and what impact there is on the natural environment. Furthermore, without understanding how a neighbourhood or a city is changing, whether the population is growing or shrinking and in which ways, what the civil engineer creates runs the risk of missing the mark. Understanding context allows civil engineers to create designs and solutions that are responsive to the particular problems and opportunities of a project, as well as fulfil the demands of the local community.

1.2 PESTLE ANALYSIS, SWOT, AND RELATED METHODS

Civil engineering and the construction and built environment fields more generally are not unique in their need to assess external factors. The business world, including fields such as human resources management and marketing, has a long history of trying to account for outside influences, appreciating the necessity of having a wider scope than any single project or initiative that the organisation might be undertaking. This wider scope enables the organisation to make strategic decisions about the likelihood of success of the project. The factors are considered to be external because they are not within the control of the organisation undertaking the analysis.

One method of analysis is termed PESTLE, which stands for Political, Economic, Sociological, Technological, Legal, and Environmental, and alternately referred to as PESTEL, with the final two factors transposed (Perera 2017). Similar methods of analysis include a variety of other acronyms, such as ETPS (Economic, Technical, Political, and Social), STEPE (Social, Technical, Economic, Political, and Ecological), STEEPLE (Social/Demographic, Technological, Economic, Environmental, Political, Legal, and Ethical), and PESTLE/PESTEL where the word Labour is substituted for Legal (Rastogi and Trivedi 2016). The earliest academic discussion of these methods dates back to the 1960s (Aguilar 1967). What all these methods have in common is the recognition of the importance of including external factors when engaging in strategic risk management. In a for-profit business setting, the rationale for engaging in these analyses is underpinned by a need to ensure that the project in question is economically viable. For civil engineers, a more wholistic view of sustainability, including whether the project meets criteria for long-term social, economic, and environmental sustainability, can be the motivator for undertaking a PESTLE or associated analysis.

PESTLE and related analytical methods can be combined with SWOT (Strengths, Weaknesses, Opportunities, and Threats) analysis (Casañ et al. 2021). In a traditional SWOT analysis, the strengths and weaknesses are internal to the organisation and the opportunities and threats are external (see Figure 1.1). Combining PESTLE and SWOT results in identifying internal strengths and weaknesses and external threats and opportunities for each of the factors (e.g., political), which are no longer seen as only external. PESTLE and SWOT together might also focus on the threats and opportunities, with each identified through the lens of the various PESTLE factors. For example, upcoming

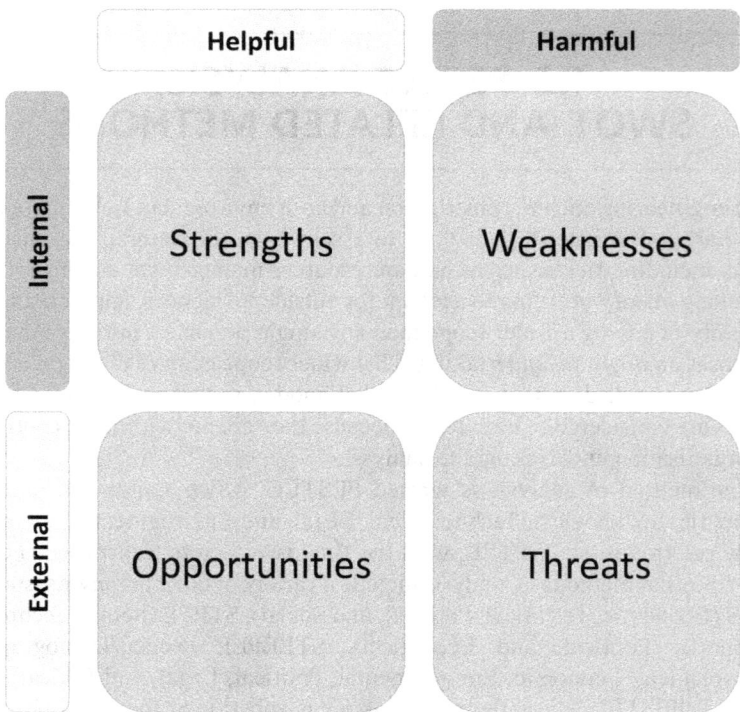

	Helpful	Harmful
Internal	Strengths	Weaknesses
External	Opportunities	Threats

FIGURE 1.1 SWOT Analysis.

changes in regulations, which fall under the 'political' category for PESTLE, might represent either an opportunity or a threat, depending on whether those changes are favourable for the project or initiative being considered.

These analytical methods are well suited to assessing construction and other civil engineering projects, but the underlying assumption of the methods is that the project manager and those involved in the analysis have an understanding of each of those factors, can accurately identify the factors at play, and can assess the relative importance of each of the factors for the given project. Whichever analytical method or tool is used (e.g., PESTLE analysis), an incorrect or incomplete understanding of the factors will inevitably lead to a flawed analysis as important factors are neglected or misjudged, or relatively unimportant factors are given unnecessary weight. However, by building an awareness of the spectrum of factors that may need to be considered, and developing an understanding of the role that context can play more generally, a civil engineer or project manager can more confidently evaluate their proposed project.

1.3 RELEVANCE OF THE UNITED NATIONS' SUSTAINABLE DEVELOPMENT GOALS

In 1987, the Secretary General of the United Nations issued a document titled Report of the World Commission on Environment and Development, which in the decades since its release has come to be known as the Brundtland Report, named after the Secretary General and lead author (Brundtland 1987). The Brundtland Report gave us the most widely accepted and frequently cited definition of sustainable development: 'development that meets the needs of the present without compromising the ability of future generations to meet their own needs' (43) (Brundtland 1987). It is also from this report that the concept of sustainability as having three pillars – economic, environmental, and social – was derived.

In the decades since the release of the Brundtland Report, the United Nations has progressed its work on sustainable development. In 2015, the UN released what is commonly referred to as the 2030 Agenda, defining 17 Sustainable Development Goals with 169 targets, to be met by 2030 and addressing the world's most pressing problems (UN 2015). The 17 SDGs have evolved away from the traditional three pillars toward the "5Ps" of people, planet, prosperity, peace, and partnership, emphasising the interconnectedness of the goals.

The 17 SDGs, which are sometimes referred to by their number, include 1) no poverty, 2) zero hunger, 3) good health and well-being, 4) quality education, 5) gender equality, 6) clean water and sanitation, 7) affordable and clean energy, 8) decent work and economic growth, 9) industry, innovation, and infrastructure, 10) reduced inequalities, 11) sustainable cities and communities, 12) responsible consumption and production, 13) climate action, 14) life below water, 15) life on land, 16) peace, justice, and strong institutions, and 17) partnerships for the goals (UN 2015).

The urgency in needing to meet these SDGs is only increasing. In November of 2022, the UN announced that the world's population had reached 8 billion people (UN 2022). The growing population places increasing strain on the built environment, economic systems, and access to resources like food and clean water, and a growing population means increased demand for housing and transportation, amongst other things (Passer et al. 2020). In meeting the needs of this growing population, built environment professionals have a professional and ethical duty to understand and mitigate the detrimental effects of the energy used and the GHG emissions released during the construction, operation, and ongoing maintenance of the built environment, as an estimated 2.5 trillion square feet of new buildings will be needed by 2060 (Abergel et al. 2017). Civil engineers and other built environment professionals furthermore must shoulder the responsibility of designing and building in a way that can protect people from the adverse effects of the climate crisis, such as more extreme temperatures and weather conditions.

In each chapter of this book, the relationship between the contextual factor and one or more SDG will be explained, demonstrating how civil engineers' understanding of context should directly inform their decisions in creating a more sustainable built environment and, by extension, a more sustainable planet.

1.4 HOW TO READ THIS BOOK

Whether you are an undergraduate or postgraduate student currently working towards a civil engineer degree, or whether you are a seasoned professional with years of experience working in the built environment, my hope is that you have opened this book because you recognise that a well-rounded understanding of the context in which engineers operate could very well make you a better engineer, one who is more responsive to the needs of their community, more aware of the history of what has occurred, and better prepared to wrestle with problems for which there is no clear solution.

You may wish to read this book from cover to cover, as it is a relatively short book, but you may instead flip to chapters that seem most relevant for whichever decision you are currently weighing. If the latter applies to you, I humbly ask that you recognise that in many circumstances, a holistic understanding of all contextual factors (presented here as chapters) would best serve the situation.

REFERENCES

Abergel, T., B. Dean., and J. Dulac. 2017. "Towards a Zero-emission, Efficient, and Resilient Buildings and Construction Sector: Global Status Report 2017." *UN Environment and International Energy Agency: Paris, France* 22.

Aguilar, F. J. 1967. *Scanning the Business Environment.* New York: Macmillan.

Brundtland, G. H. 1987. *Report of the World Commission on Environment and Development.* New York: Oxford University Press.

Casañ, M. J., M. Alier, and A. Llorens. 2021. "A Collaborative Learning Activity to Analyze the Sustainability of an Innovation using PESTLE." *Sustainability* 13(16): 8756.

Passer, A., T. Lützkendorf, G. Habert, H. Kromp-Kolb, M. Monsberger, M. Eder, and B. Truger. 2020. "Sustainable Built Environment: Transition Towards a Net Zero Carbon Built Environment." *The International Journal of Life Cycle Assessment* 25(6): 1160–1167.

Perera, R. 2017. *The PESTLE Analysis.* Avissawella: Nerdynaut.

Rastogi, N., and M. Trivedi. 2016. "PESTLE Technique – A Tool to Identify External Risks in Construction Projects." *International Research Journal of Engineering and Technology (IRJET)* 3(1): 384–388.

UN. 2015. *Transforming Our World: The 2030 Agenda for Sustainable Development.* New York: United Nations, Department of Economic and Social Affairs.

UN. 2022. "Day of 8 Billion." United Nations. (n.d.). "Day of 8 Billion." Retrieved from https://www.un.org/en/dayof8billion.

Local and Regional Politics

2

NIMBYism and beyond

2.1 CONNECTING THE UNITED NATIONS' SUSTAINABLE DEVELOPMENT GOALS TO POLITICS

The Sustainable Development Goals (SDGs) and their 169 targets, as well as Agenda 2030 more broadly, are inherently political, as they were created as a governance tool to guide public policies toward sustainable development (Biermann, Hickmann et al. 2022). The adoption of the SDGs by nation–states around the globe was, furthermore, an act of international cooperation in setting sustainability goals and targets, which, although there was a similar idea behind the Millennium Development Goals, is still an important shift away from single issue-focused policymaking and toward a recognition of the complexity, and cross-cutting nature, of most aspects of sustainability (Bornemann and Weiland 2021).

Even though the SDGs are intrinsically political, whether the SDGs have accomplished changing the nature of the political handling of sustainability is a separate issue. A meta-analysis of more than 3,000 scientific studies on the SDGs, published between 2016 and 2021, examined three areas of potential effects: discursive (changing the way sustainability is discussed, including whether the SDGs are explicitly referenced), normative (changing legislation, policies, or regulations to be brought into alignment with the SDGs), and

DOI: 10.1201/9781003392194-2

institutional (the creation of new departments, programmes, or other sources of policymaking for the express purpose of achieving the SDGs) (Biermann, Hickmann et al. 2022). The authors of that study argued that transformational change would be achieved only with all three kinds of effects in place, but they concluded that most changes have only been discursive (Biermann, Hickmann et al. 2022).

Even if the results of the meta-analysis are correct and that the impact of nation–states adopting the SDGs has not included real change in the way that infrastructure is being built or the kinds of infrastructure being built, an argument can be made for the burden being, at least partially, on the shoulders of civil engineers and related professionals. Policy makers have indicated that the SDGs matter; they are making discursive reference to them. Yet these same policy makers may lack the know-how for how to actually achieve those SDGs. Engineers, on the other hand, have the knowledge, expertise, and responsibility of designing in a way that will help the world meet those SDGs (Addagada, Ramesh et al. 2022). Indeed, the United Nations as well as the global Engineers Without Borders challenge both use the term 'global responsibility' to refer to how engineers need to consider a holistic view – the economic, social, and environmental sustainability – of their work (Chance, Direito et al. 2022).

Many engineers may prefer to stay out of politics, but civil engineering is inherently a people-serving profession, and ensuring that the safety, welfare, and needs of the populace are being met may mean working with politicians, especially ones in local government (Wiewiora 2005). With the adoption of the SDGs, the role of the civil engineer becomes even more political, as local politicians may be unable to sift through the overwhelming data, which is often technical, to properly guide their decision making, and may rely more on the advice of the civil engineer for what should be built, where, and how. Furthermore, even if the politicians are not seeking out this advice or input, and especially if they are resistant to receiving it, engineers may need to take it upon themselves to promote designs and to shape solutions that are in the best interests of the people – and in the best interest of meeting the SDGs.

2.2 UNDERSTANDING THE LOCAL POLITICAL CLIMATE

The structure of government can have a direct influence on the local political climate. In the United Kingdom, the United States, and France, there are rough corollaries of sub-national governments at the regional level: counties, states, and departments, respectively. At the local level, there is the parish (UK), township (US), and commune (France) for rural areas, and of course all three countries additionally have cities and towns. The decision-making powers for each of these governmental entities are derived from various statutes and charters, which has allowed for something of a divorce between federal and local politics in many instances (Chandler 2013).

In the US, the two most common structures for local government are the council–manager (CM) and mayor–council (MC) structures (DeSantis and Renner 2002). One can think of the CM structure as analogous to a business, where the manager serves in a role similar to a CEO, answerable to a board of a directors (the council) and shareholders (citizen voters); the council hires, and can fire, the manager (Hayes and Chang 1990). When there is a mayor within the CM structure, it is a figurehead appointment without formal powers. Within the MC structure, on the other hand, the mayor is a popularly elected official and the council serves as the legislative branch. At the risk of oversimplification, the additional important distinction between these structures lies primarily in the authority and executive powers of the manager or mayor, where the manager largely shares authoritative power and executive responsibility with the council, whereas the mayor predominantly does not (Carr 2015). One practical result of these differences is that the mayor can gain fame or blame for decisions made, and may win or lose the next election based on how popular his or her decisions are amongst the electorate, but the manager is more likely to work in tandem with the council, to whom he or she is accountable.

In the UK, the structure of local government is more complicated, owing to a series of reforms and restructuring over decades, if not centuries. Whilst Northern Ireland has elected borough, city, and district councils, Wales and Scotland have unitary authorities, sometimes termed county councils, as well as community councils that are analogous to parish and town councils in England. In England, there are county councils, district councils (which are divisions of county councils), unitary authorities (a single tier of local government for some towns, cities, and small counties), plus metropolitan districts and the unitary authority of each London borough. Additionally, there are town and parish councils in various places in England, and in a small number of places there is a mayor for ceremonial purposes only (Sandford 2002).

Although countless additional pages could be written about the structure of local government in various countries, the takeaway message is that to understand the local politics, one must first understand the local political structure because that will shed light on how influential the local politicians are likely to be and, more importantly, how much power they hold in the decision-making process. The local mayor, council member, or other politician might hold sway over public opinion, or they may be beholden to what the public wants.

In addition to the political structure, political influence and power can rest formally or informally in the hands of other organisations, groups, and individuals, especially private businesses, religious and educational institutions, and other longstanding stakeholders. In Britain and throughout Europe, as well as in places like the United States and Australia, the movement over the past several decades has been toward community partnerships and collaborative involvement of key stakeholders in making decisions about local development (Healey 1998). This change toward partnership and the recognition of the value of stakeholders' knowledge came from a rise in scepticism about technocratic decision making, which had dominated the built environment until at least the 1960s and is still prevalent in many places (Pacchi 2018). Around this time and for various reasons including the Civil Rights Movement, a backlash arose against so-called top-down approaches to making decisions about what would be built and where.

Civil engineers and other built environment professionals may believe that their technical expertise in designing and building, for example, an airport expansion is all that is needed for the project to move forward. After all, engineers spend years in school studying technical material to ensure that their structures are safe and sound and, depending on their experience, can be experts in recommending appropriate building locations, as they are equipped to evaluate the soil conditions, for example, the choice of building materials, the overall massing, and even innovations in design above and beyond what the architect may have recommended. However, increasingly the expectation is that the community should be actively involved and their feedback should be solicited, which can sometimes lead to an unhappy tension between the various parties – between the technical experts and the community members, who may indeed be local-knowledge experts and influential stakeholders (Schatz and Rogers 2016). Even when the engineer is not directly involved in facilitating community participation, it would benefit him or her to seek out this contextual information – who the key players are within the community and what strong opinions they may have about what should, or should not, be built.

2.3 NIMBY, BANANA, AND THE ALPHABET SOUP

Built environment professionals commonly encounter resistance to proposed projects because of concerns about the proposed land use itself, an increase in nuisance (such as noise, pollution, traffic, or smells), the impact on property values, or changes to the character of the neighbourhood. These concerns may or may not be justified, but what is important is that those who have the concerns perceive them to be true or at least possible. This resistance is often described by one of many acronyms that describe relatively similar concepts: BANANA (built absolutely nothing anywhere near anyone/anything), CAVE (citizens against virtually everything), LULU (locally undesirable/unwanted land use), NAMBI (not against my business or industry), NIABY (not in anyone's backyard), NIMBY (not in my backyard), NIMTOO (not in my term of office), and NOPE (not on planet Earth), amongst others (Schively 2007). If these acronyms have a negative connotation, it may be because they express built environment professionals' frustration with trying to build something that is needed (e.g., a power plant, a wastewater treatment facility, a bus depot, an affordable-housing complex, a new road), only to meet resistance from the public, elected officials, and/or other stakeholders such as business owners.

LULUs were first termed as such in the early 1980s, with the recognition that they represent a particularly difficult problem since they include almost every major development project (Popper 1981). They are land uses that are needed in communities but are also negatively regarded by community stakeholders. For example, an airport represents a significant long-term investment and, in some cases, can be a welcome enhancement to the transportation network (Caves and Gosling 1997). However, airports are noisy – both in the surrounding area and in the flight paths to and from the airport – and can cause considerable environmental impacts, and thus proposed airports and airport expansions can encounter significant opposition from environmental groups (Popper 1985).

The Bristol (UK) airport expansion project offers one such example. In late 2018, Bristol Airport Limited submitted a planning application for expansion, in which they describe their proposed changes to allow for an increase from the current demand of 9 million terminal passengers per annum to a proposed 12 million (Bristol Airport Limited 2018). Their application reviews the history of the airport, which was opened in 1957 and has been expanded several times, most recently in 2011 to allow for an operational capacity of 10 million terminal passengers per annum (Bristol Airport Limited 2018). The proposal submitted in 2018 sought planning permission for extensive physical changes – including terminal expansions, a new walkway, a new pier, a new multistorey carpark, and road expansions, amongst other construction – in addition to asking for the removal of the seasonal restrictions on night flights. The document identified 24 planning issues, including greenhouse gas emissions/climate change, noise impacts, air quality, impact on the Green Belt, biodiversity, flood risk and drainage, and land quality, clearly signalling Bristol Airport's awareness that there would likely be objectors to their proposed expansion. Indeed, filed with the proposal is an additional document from 2020 that notes concerns from the Cardiff International Airport, Parish Councils Airport Association, and Bristol Airport Action Network (BAAN) (North Somerset Council 2020). Whilst the Bristol Airport maintains that airport expansion would bring economic development to the area and that potential impacts on the environment have been fully taken into consideration, objectors to the expansion, most notably BAAN, argue that the economic benefits would be marginal and that the environmental concerns such as the increased air and noise pollution and increased greenhouse gas emissions, as well as destruction of parts of the Green Belt, are significant and will make it impossible to meet local, regional, and national targets for carbon reduction (BAAN 2022). The airport expansion project continues to be in a state of limbo as the council decisions are appealed. This case illustrates the complexity of viewpoints involved, and the political tension that can arise, and should make the reader aware that planning an airport expansion goes beyond the design of the expansion itself.

The NIMBY phenomenon is related, as sometimes the objection to a LULU is not as black and white as should it be built or not, but rather the objection is to the location – specifically, the objectors do not want the development to take place near their residence, business, or other such property. There may be recognition that the development would bring benefits to the region or would serve a regional or societal need, but the objection is to the development being sited locally. The NIMBY phenomenon is so pervasive that built-environment professionals sometimes refer to individuals and groups as NIMBYs, and their state of mind as NIMBYism. Justifiably or not, they can be seen as opponents of development and a hindrance to progress.

In addition to airports, other development projects such as landfills, wastewater treatment facilities, prisons, power plants, wind farms, and utility lines are often subject to a NIMBY response. Whilst the objections may seem to be politically agnostic, the NIMBY reaction is also provoked by proposals for projects such as low-income housing, multifamily housing, drug and rehabilitation centres, and bars. Objections to these developments may be based on facts (such as a projected increase in demand for parking), on perceptions and fears (such as an increase in crime), or a combination. The situation becomes more complicated as issues of socioeconomic class and sometimes race, ethnicity, and immigration and refugee status enter into the equation.

What can be done to overcome the NIMBY response? The key is to plan ahead. When projects are still in their conceptual phase, it is essential to involve the local community in well-mediated, collaborative processes. (For readers in the UK, the recommendation is that this involvement begins no later than RIBA Stage 2 (RIBA 2022)). This is important for several reasons. First, in places where the democratic process is a right, it allows for substantial input and influence from the community. Second, by involving the stakeholders from the beginning, they are more likely to have buy-in for the project, to see themselves as having had a hand in crafting the development. And third, this allows for those built-environment professionals involved in the project to educate the community about the facts of the project and its impact, and to address perceptual concerns at an early stage in the process.

The terms CAVE and BANANA are used solely as pejoratives, including referring to objectors to a proposal as 'CAVE people.' The pejorative language expresses the frustration that built-environment professionals may have when it seems that any and all proposals are met with resistance from a vocal public. While the NIMBY reaction is provoked in response to the location, it allows for the acknowledgement of the merits of the development. For example, most people likely do not want to live next to a power plant, but they will reasonably acknowledge the need for a power plant of some kind to be located within the service area. With CAVE, BANANA, NIABY, and NOPE resistance, the objections might be rooted in the land use itself, such as objecting to a new nuclear power station because of an opposition to nuclear power as an energy source. Some of these objectors might be convinced of the merits of a given project, if presented with compelling facts and arguments, whereas others will inevitably not be swayed.

A common reason given for objections to development is that the proposed development will change the character of the neighbourhood or area and its visual, aesthetic, or architectural traditions. A tall residential tower in a neighbourhood with four- and five-storey buildings could meet with various objections: the tower might block daylight from reaching the neighbouring buildings, the increase in density will increase vehicular traffic and demand for parking, and the scale and massing of the building is incongruent with the neighbourhood. The issues of daylight/overshadowing and traffic/parking can be addressed through the design of the project, but the incongruence of the building with the neighbourhood is a battle that is harder to fight. The strategy is the same as with NIMBY responses, where the key is to involve stakeholders early in the process. Although this can give staunch opponents more time to develop their case against the proposal, it also opens up the possibility of adequately addressing their objections and alleviating their concerns.

2.4 THE URBAN AND RURAL DIVIDE

In westernised democracies, and especially in the United States, there exists a political divide between urban and rural areas, to the extent that urban areas tend to be more liberal and the rural areas tend to be more conservative. This political divide has roots going back to at least the Industrial Revolution, with scholars speculating that the rural residents, with their agriculture-based economy, had different political interests from urban residents, some of whom were benefitting from the rapid industrial, technological, and economic changes that also brought about social change (Kenny and Luca 2021). This divide has continued to exist, although scholars disagree about the extent to which the divide has changed or even faded, and whether other demographic characteristics such as educational attainment have become more influential in political orientation (Kenny and Luca 2021).

In contemporary westernised democracies, it can be useful to consider the urban–rural divide, but also to view it as more nuanced, as more of a continuum than a dichotomy. This is especially true in so-called 'swing states' in the US, which tend to have more of a gradient of voting patterns represented, rather than a division that is strictly left–right (liberal–conservative, Democratic–Republican, blue–red) (Scala and Johnson 2017). In the UK, especially within England, the urban–rural divide is overlaid with economic growth/decline: there are metropolitan areas experiencing growth and attracting a highly paid, highly skilled workforce (e.g., Cambridge), and there are large swaths of areas, including some cities, that are experiencing decline and unable to attract those aforementioned workers (e.g., coastal areas in the northeast of England) (Jennings and Stoker 2016). This is not dissimilar to some areas of the US, such as in the rust belt and its shrinking cities. This urban–rural divide is present in continental Europe as well, especially in Western Europe (Kenny and Luca 2021).

An ongoing issue with regard to the divide, whether viewed in stark terms or as a continuum, is that the political considerations are different. Limiting the discussion to the built environment, there is still the issue that many rural inhabitants feel 'left behind,' believe that money is invested in the booming cities rather than locally, and furthermore are less likely to trust their government, possibly extending as far as to the kinds of projects being built (Kenny and Luca 2021). In the booming cities, the political considerations are more likely to concern competition over scarce housing supply, whether transport links are fast, abundant, and affordable, and whether government actions are environmentally responsible (Jennings and Stoker 2016).

Civil engineers working on projects that seem to position rural areas as backwater, flyover country, or in any other terms that make them seem less important than their cosmopolitan counterparts, should take caution in how these projects will be perceived, and should expect to encounter resistance. For example, the politics of water allocation can be particularly contentious, as metropolitan areas need increasingly large supplies of clean water and the associated construction of dams and reservoirs can leave rural areas without adequate, reliable, or clean water (Punjabi and Johnson 2019).

2.5 STATE AND PROVINCIAL GOVERNMENTS AND THEIR INFLUENCE

In countries such as the United Kingdom, United States, Canada, France, Kenya, and Malaysia, amongst others, a political tension has existed between federal powers and decentralised powers, which is to say powers allocated instead to the states or provinces or to local government. In Canada and the UK, this has been termed the devolution of government, with the transfer of power, responsibility, and funding to sub-federal levels (Lonti 2005). In the UK, the devolution of government has been asymmetric, in the sense that it has occurred to differing degrees within different parts of the UK. In the US, this tension is often framed in terms of states' rights versus federal power, with Democrats favouring the national government having a strong role in regulating the economy and Republicans favouring a so-called free market approach where, if government is to intervene, the intervention or regulation takes place at the state level.

The reasons for decentralisation and devolution are as varied as the number of countries and, in certain cases, can be potentially problematic. In some nations, at various points in history, decentralisation has been combined with a reworking of internal boundaries in a way to favour one group of people over another, such as white Americans over African Americans (McGarry 2007). This has happened through ensuring that one demographic group remains the majority in a state, province, or region either because it is the group in power, as a means of isolating a minority group, or as a way of empowering a minority group by creating a minority–majority area. In Canada, when the province of Quebec has periodically pushed for more autonomy, the federal government has reacted by extending that autonomy to each region, creating a largely decentralised government (McGarry 2007).

The practical effect of decentralised government is that, in some places, the level of government that has the most power and influence on building works may very well be something other than the federal government. For example, in the United States, the National Environmental Policy Act (NEPA) applies to projects that receive federal funding, work, or permits, requiring that any potential significantly negative environmental impacts are disclosed before work is begun, but certain states have passed additional legislation. For example, the California Environmental Quality Act (CEQA) pertains to state and local agencies' building works but goes a step further than NEPA in that the significant environmental impacts may be required to be avoided in order to receive permitting for the project (O'Brien 2009). Whilst this is a legal consideration and not directly a political one, it is worth considering how differences in state legislation may affect the political climate of the area.

Separate from the legal–political considerations are the economic–political considerations, which can go hand in hand with private companies. Take the example of the multinational energy corporation RWE. RWE is headquartered in Germany and operates through Europe, including the UK, as well as in Asia and the United States. Although RWE has a portfolio of energy sources, including natural gas, nuclear power, and renewables (including offshore wind/solar and hydro), the company has been heavily criticised for most of the 21st century because of its open-pit lignite mining in the Hambach Forest, in the German state of North Rhine-Westphalia (NRW). On the surface, there appears to be a standard public–private partnership where RWE is the majority supplier of power in the state. However, RWE's political influence in the state cannot be overexaggerated. Any proposed energy project in NRW would certainly be noticed by RWE, if not outright opposed, and engineering firms would need to understand the extent of RWE's influence on local politicians as well (Brock and Dunlap 2018).

2.6 INDIGENOUS GOVERNANCE

As of 2018, Indigenous peoples managed or had long-term recognised rights of occupation to over a quarter of the world's land, which equates to at least 38 million square kilometres within about 90 nation–states; furthermore, this represents a high proportion of the world's conservation areas (Garnett et al. 2018). The term 'Indigenous peoples' is sometimes replaced by other terms, such as First Nation or Native American, but as an umbrella term refers to people who 'claim ancestry from a self-governing society that inhabited a region before the invasion, conquest, settlement, or other form of occupation by people of different cultures who then became dominant' (Muckle 2012). There are Indigenous peoples on every continent besides Antarctica.

The degree to which Indigenous peoples have autonomous governance over their ancestral lands varies tremendously but almost always includes a complicated and contentious history and sometimes present, where Indigenous communities are not recognised as political entities by nation–states (Champagne 2011). For example, the boundaries of the United States include more than 500 sovereign Native American Tribes, which are legally categorised as domestic dependent nations – meaning that their sovereignty is limited to domestic and internal affairs (Krakoff 2004). A considerable amount of conflict has ensued from the issue of the right to manage drinking water and other resources (e.g., natural resources such as uranium). More than a century has passed since the Winters Doctrine, which guaranteed that Reservation land would permanently belong to each Tribe and, with that land, came the rights to adequate water supplies to meet the needs of those living on the Reservation. In some cases, such in the arid Southwestern United States, Tribes have entered into legal relationships with non-Native communities to lease water rights on a temporary and for-profit basis, but the practice has been controversial, especially because of the history of non-Native water users taking the Tribal water for free (Jones and Ingram 2022).

For the Navajo (Diné) Nation, which includes more than 300,000 members, half of whom live on the Reservation, the water issue is further complicated by the presence of more than 500 deserted uranium mines, which have contaminated the water supply (Rock and Ingram 2020, Jones and Ingram 2022). Various US federal agencies and policies have sought to address the uranium contamination issue, but these approaches have been largely culturally inappropriate. The Navajo Nation specifically has its own set of Fundamental Laws – traditional laws that guide how Diné's relationship with animals, people, land, and the natural environment progresses (Rock and Ingram 2020).

These Fundamental Laws represent traditional ecological knowledge that needs to be respected, especially if one is considering any civil engineering work that would impact Navajo land. Furthermore, because the intricacies of the Fundamental Laws are known only to Navajo elders, an engineer from outside the Tribe cannot expect to dictate solutions (Rock and Ingram 2020). Even where sovereignty is largely denied, such as with the Māori people of Aotearoa New Zealand, the importance of respect for traditions, values, laws, desires, and goals of the Tribe or Nation itself cannot be overstated (O'Brien 2019).

REFERENCES

Addagada, L., S. T. Ramesh, D. N. Ratha, R. Gandhimathi, and P. R. Rout. 2022. *The Role of Civil Engineering in Achieving UN Sustainable Development Goals*, 373–389. Singapore: Springer.

BAAN. 2022. *Why Oppose Airport Expansion?* www.stopbristolairportexpansion.org/why-oppose-airport-expansion/.

Biermann, F., T. Hickmann, C. A. Sénit, M. Beisheim, S. Bernstein, P. Chasek, L. Grob, R. E. Kim, L. J. Kotzé, and M. Nilsson. 2022. "Scientific Evidence on the Political Impact of the Sustainable Development Goals." *Nature Sustainability* 5(9): 795–800.

Bornemann, B., and S. Weiland. 2021. "The UN 2030 Agenda and the Quest for Policy Integration: A Literature Review." *Politics and Governance* 9(1): 96–107.

Bristol Airport Limited. 2018. *Planning Application No. 18/P/5118/OUT – Outline Planning Application for the Development of Bristol Airport*. North Somerset Council. Retrieved from https://www.n-somerset.gov.uk/sites/default/files/2020-09/Notice%20of%20Decision%20Bristol%20Airport%20Ltd.pdf

Brock, A., and A. Dunlap. 2018. "Normalising Corporate Counterinsurgency: Engineering Consent, Managing Resistance and Greening Destruction around the Hambach Coal Mine and Beyond." *Political Geography* 62: 33–47.

Carr, J. B. 2015. "What Have We Learned about the Performance of Council-manager Government? A Review and Synthesis of the Research." *Public Administration Review* 75(5): 673–689.

Caves, R. E., and G. D. Gosling. 1997. *Strategic Airport Planning*. Bingley, UK: Emerald Group Publishing Limited.

Champagne, D. 2011. "Indigenous Nations of the World." *Indian Country Today*. Published Nov 17, 2011. Updated Sept 13, 2018.

Chance, S., I. Direito, and J. Mitchell. 2022. "Opportunities and Barriers Faced by Early-career Civil Engineers Enacting Global Responsibility." *European Journal of Engineering Education* 47(1): 164–192.

Chandler, J. A. 2013. "Explaining Local Government: Local Government in Britain since 1800." In *Explaining Local Government*. Manchester: Manchester University Press.

DeSantis, V. S., and T. Renner. 2002. "City Government Structures: An Attempt at Clarification." *State and Local Government Review* 34(2): 95–104.

Garnett, S. T., N. D. Burgess, J. E. Fa, Á. Fernández-Llamazares, Z. Molnár, C. J. Robinson, J. E. Watson, K. K. Zander, B. Austin, and E. S. Brondizio. 2018. "A Spatial Overview of the Global Importance of Indigenous Lands for Conservation." *Nature Sustainability* 1(7): 369–374.

Hayes, K., and S. Chang. 1990. "The Relative Efficiency of City Manager and Mayor-council Forms of Government." *Southern Economic Journal*: 167–177.

Healey, P. 1998. "Building Institutional Capacity through Collaborative Approaches to Urban Planning." *Environment and Planning A* 30(9): 1531–1546.

Jennings, W., and G. Stoker. 2016. "The Bifurcation of Politics: Two Englands." *The Political Quarterly* 87(3): 372–382.

Jones, L., and J. C. Ingram. 2022. "Invited Perspective: Tribal Water Issues Exemplified by the Navajo Nation." *Environmental Health Perspectives* 130(12): 121301.

Kenny, M., and D. Luca. 2021. "The Urban-rural Polarisation of Political Disenchantment: An Investigation of Social and Political Attitudes in 30 European Countries." *Cambridge Journal of Regions, Economy and Society* 14(3): 565–582.

Krakoff, S. 2004. "A Narrative of Sovereignty: Illuminating the Paradox of the Domestic Dependent Nation." *Oregon Law Review* 83: 1109.

Lonti, Z. 2005. "How Much Decentralization?:Managerial Autonomy in the Canadian Public Service." *The American Review of Public Administration* 35(2): 122–136.

McGarry, J. 2007. "Asymmetry in Federations, Federacies and Unitary States." *Ethnopolitics* 6(1): 105–116.

Muckle, R. J. 2012. *Indigenous Peoples of North America: A Concise Anthropological Overview*. Toronto: University of Toronto Press.

North Somerset Council. 2020. *Planning and Regulatory Committee Update Sheet*. Published by North Somerset Council. Retrieved from https://n-somerset.moderngov.co.uk/Data/Planning%20and%20Regulatory%20Committee/202002101800/Agenda/04.1%20Update%20Sheet%2018p5118out%20Bristol%20Aiport.pdf.

O'Brien, C. 2009. "I Wish They All Could Be California Environmental Quality Acts: Rethinking NEPA in Light of Climate Change." *Boston College Environmental Affairs Law Review* 36: 239.

O'Brien, T. 2019. "Political Entrepreneurship in the Field of Māori Sovereignty in Aotearoa New Zealand." *The British Journal of Sociology* 70(4): 1179–1197.

Pacchi, C. 2018. *Critical Voices from the 1960s: Jane Jacobs and the Epistemological Critiques to the Technocratic Planning Model* (Jane Jacobs 100: her legacy and relevance in the 21st century). Delft: TU Delft BK Bouwkunde.

Popper, F. J. 1981. "Siting LULUs (Locally Unwanted Land Uses)." *Planning (ASPO)* 47(4): 12–15.

Popper, F. J. 1985. "The Environmentalist and the LULU." *Environment: Science and Policy for Sustainable Development* 27(2): 7–40.

Punjabi, B., and C. A. Johnson. 2019. "The Politics of Rural – urban Water Conflict in India: Untapping the Power of Institutional Reform." *World Development* 120: 182–192.

RIBA. 2022. *RIBA Plan of Work*. www.architecture.com/knowledge-and-resources/resources-landing-page/riba-plan-of-work.

Rock, T., and J. C. Ingram. 2020. "Traditional Ecological Knowledge Policy Considerations for Abandoned Uranium Mines on Navajo Nation." *Human Biology* 92(1): 19.

Sandford, M. 2002. *Local Government in England: Structures*. London: House of Commons Library.

Scala, D. J., and K. M. Johnson. 2017. "Political Polarization along the Rural-urban Continuum? The Geography of the Presidential Vote, 2000–2016." *The ANNALS of the American Academy of Political and Social Science* 672(1): 162–184.

Schatz, L., and D. Rogers. 2016. "Participatory, Technocratic and Neoliberal Planning: An Untenable Planning Governance Ménage à Trois." *Australian Planner* 53(1): 37–45.

Schively, C. 2007. "Understanding the NIMBY and LULU Phenomena: Reassessing our Knowledge Base and Informing Future Research." *Journal of Planning Literature* 21(3): 255–266.

Wiewiora, J. A. 2005. "Involvement of Civil Engineers in Politics." *Journal of Professional Issues in Engineering Education and Practice* 131(2): 102–104.

National Politics
Who Cares about the Climate Crisis?

3

3.1 THE ROLE OF NATIONAL POLITICS IN LOCAL ACTIONS

This section will consider two perspectives in how national politics can influence the ability of engineers to engage in meaningful local action: how the climate policies of a nation relate to the Paris Agreement, and how the political pendulum of a nation can change the actions available.

3.1.1 The Paris Agreement

The Paris Agreement is an international treaty on climate change where nations have agreed to set their own targets, or Nationally Determined Contributions (NDCs), for reducing their carbon emissions, with the goal of mitigating and adapting to the effects of climate change. The Paris Agreement was signed by representatives from virtually all nations, and ratified by all but four; of those four, the only notable high emitter is Iran (Abdollahi 2021). Whilst the Paris Agreement represents a pledge to reduce carbon emissions, the Sustainable Development Goals offer the goals and targets that nations can use to influence their policies.

The NDCs are bottom-up climate actions plans where countries were given autonomy in deciding their contributions – with the key word being contribution rather than commitment, in the sense that it is not binding. Furthermore, under the Paris Agreement, there was widespread recognition that lesser developed countries would likely require assistance in achieving their goals and targets, and thus were allowed to make their climate action plans conditional upon receiving support from more developed countries, specifically support in the form of finance, technology transfer, and/or capacity building (Pauw et al. 2020). Although the allowance of conditional contributions was made in the interest of fairness and equity amongst nations of differing development, the consequence is that if lesser developed countries do not receive the support needed to realise their contributions, no country is ultimately held accountable for that failure.

Engineering arguably offers the world's best chance at finding solutions to mitigate and adapt to climate change. Within civil engineering, there have been advances in mitigation through the development of new materials such as green concrete (Stewart 2023). There have also been advances in adaptation through designs that reduce the vulnerability of infrastructure to climate change (Stewart 2023). Within engineering, specialists in climate geoengineering have explored other mitigation and adaptation efforts that are largely outside the scope of civil engineering, such as large-scale afforestation, ocean iron fertilisation, and space mirrors (Lawrence et al. 2018). The ability to implement any of these technologies, however, relies upon supportive policy, including funding for the initiatives. Furthermore, these technologies may have the greatest positive impact in countries that are least able to afford them.

3.1.2 The Political Pendulum and Changing Priorities

The United States initially signed onto the Paris Agreement under President Barack Obama, a Democrat, in 2016. Perhaps no greater example of how the national political pendulum can swing was the 2017 announcement of President Donald Trump, a Republican, of his decision to withdraw the United States from the Paris Agreement – a decision later reversed when President Joe Biden, a Democrat, re-joined the Paris Agreement on his first day in office. However, this was not the first instance of the political pendulum of the United States swinging with regard to the climate crisis. In 2001, under President George W. Bush, another Republican, the United States withdrew from the Kyoto Protocol, which had been joined a few years prior by President Bill Clinton, a Democrat. The political pendulum in the United States frequently swings from Republican (right) to Democrat (left) and back again, with the effect of an instability of political initiative since support to combat the climate crisis is never long lived (Kenfack 2022). This can lead to uncertainty for which pursuits (e.g., renewable energy) will be supported by legislation, which creates difficulty in assessing the feasibility of more progressive projects.

To take a very different example, we can look at environmental policies and practices in China. As the world's most populous nation and with one of the largest economies, China's environmental policies and practices have a significant impact on the globe. China faces obstacles to reduce the environmental damage brought on by its rapid urbanisation and economic growth. Air pollution, especially in metropolitan areas, is one of China's main environmental problems. The government has responded by putting in place a number of policies, including encouraging the use or renewable energy and enforcing stringent emissions regulations for factories and cars. These initiatives have significantly improved the air quality of certain areas but many other areas of the country still have severe health and environmental concerns related to air pollution. China has also invested in solar and wind power, and has promised to reach carbon neutrality by 2060, yet coal remains the nation's main source of energy and China is the largest consumer of coal worldwide. China has also continued to build coal-fired power plants in other countries. Overall, even though China has advanced certain environmental policies in recent years, much more has to be done to lessen the negative environmental effects of rapid growth.

3.2 WORKING ON BEHALF OF NATIONAL INTERESTS

Politicians are charged with defining national interests. In addition to responding to the climate crisis, politicians at the national level must also advocate for national interests such as creating greater energy self-sufficiency/security, reducing energy costs, and ensuring sufficient employment opportunities. These national interests often align with other Sustainable Development Goals, such as 1. No poverty, 2. Zero hunger, and 8. Decent work and economic growth, amongst the other SDGs. The SDGs are loosely categorised into five areas: people, planet, prosperity, peace, and partnerships, and whilst each category is important and all SDGs are meant to be achieved, an individual politician's or government's agenda may more heavily favour one issue area more than others (Morton et al. 2017). For example, a politician whose chief concern is the economy, and who believes in a free market approach, may be less likely to support environmental regulations or restrictions on coal mining or fracking.

The nation of Brazil, for example, has abundant natural resources and biodiversity, but also has issues with social inequality, environmental degradation, and economic growth. Brazilian efforts to stop deforestation in the Amazon, an unparalleled natural environment that offers numerous ecological resources, have achieved major wins in recent years. The Brazilian government has passed regulations and initiated programmes to encourage sustainable land use, safeguard the rights of indigenous peoples, and support law enforcement efforts in combatting illegal logging and land grabbing; the rate of deforestation has significantly decreased as a result of these efforts. Brazil does, however, have issues with sustainable growth, particularly with regard to infrastructure, energy, and agriculture. The nation is a significant producer and exporter of commodities like soybeans, cattle, and other foods that are wrapped up with issues of deforestation, water pollution, and greenhouse gas emissions. Brazil also generates a significant portion of its electricity through hydropower, which can have negative effects on freshwater environments and indigenous communities. Brazil is additionally dealing with issues of poverty, inequality, and lack of access to essential services like healthcare and education, which may hinder the nation's ability to put sustainable development plans into place. Brazil's national interests are complicated and multidimensional, necessitating a balance between environmental conservation, social progress, and economic growth. The country has made strides in tackling deforestation but more work is needed to support sustainable development that addresses social and economic issues.

Each nation's governing structure determines the extent to which a president, prime minister, member of parliament, or other political figure has the power to influence how funds are allocated, how projects are supported, and whether laws and regulations make initiatives easier or more difficult to pursue. Engineers should be mindful of the governing structure in which they are working but also how the projects that they are designing and building advance a national political agenda.

3.3 INDIVIDUALISTIC VERSUS COLLECTIVISTIC INTERESTS

The climate crisis is global in nature. Decisions made in one nation, such as to allow the burning of fossil fuels, have consequences around the world. The Paris Agreement and the SDGs are an explicit acknowledgment that all nations must act in order to protect people everywhere and the planet as

Collectivistic   Individualistic

Left ←——————————|——————————→ Right
      Group Goals   Personal Goals
      Common Good   Individual Freedoms
      Societal Harmony   Personal Expression

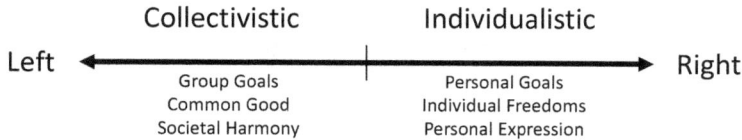

FIGURE 3.1   Collectivistic–Individualistic Spectrum.

a whole. However, within an individual country, the priority of competing national interests may well be tied to where that country falls on the individualistic to collectivistic spectrum. Whilst all national interests are, in a sense, collectivistic in that they are meant to provide for the greater good of the country, researchers have begun asking questions about whether there is a link between individualism–collectivism and climate change denial as well as climate change inaction (Xiang et al. 2019, Nartova-Bochaver et al. 2022).

In brief, all people, within groups and as individuals, fall somewhere on the spectrum of individualistic to collectivistic (see Figure 3.1). People in individualistic cultures tend to give priority to their personal goals over the goals of the group (or nation), whereas people in collectivistic cultures tend to put the group's goals above their own personal goals (Triandis 2001). In individualistic societies, people may view environmental issues through the lens of how great of an impact it has on their personal life, and environmental policies and laws may therefore be less harsh or strictly enforced. On the other hand, in collectivistic societies, people may view addressing environmental issues as a community obligation, and thus environmental policies and laws may be stronger and more strictly enforced. People in collectivistic societies may be more receptive to government laws and interventions aimed at protecting the environment, as well as more willing to accept individual sacrifices for the common good. People in individualistic societies may be more wary of government laws and interventions, as they may be viewed as an intrusion on their own freedoms.

Additionally, more individualistic societies may view consumerism and its effects on the environment differently than more collectivistic societies. Material goods may be considered as a means of expressing one's personality and aspirations in individualistic societies, where one's personal freedoms and individual choices are highly prized, often with little regard to how consumption may affect the environment or society as a whole. People in these societies could be more inclined to regard environmental issues as something for which individuals should decide for themselves how to respond, as opposed to needing group action. In more collectivistic societies, consumerism may be viewed as a danger to sustainability and community well-being, where social responsibility is highly valued. It may be easier to convince people to consume less and put sustainability before personal preferences. For example, there is a Japanese environmentalist term 'mottainai' (roughly, 'What a waste!' or 'Waste not, want not'), which conveys the idea of a respect for resources and thriftiness, which in turn can be seen in high rates of recycling and energy saving. In collectivistic societies, the long-term, negative effects of consumerism on the environment and on generations to come may be emphasised over the short-term gains for the individual. This is evident in nations like Bhutan, which embraced the idea of a Gross National Happiness (GNH) metric of well-being that emphasises society harmony and environmental sustainability over economic gains.

Individualistic countries include Australia, the United States, and much of Western Europe, whereas collectivistic countries include most of Africa and Asia as well as parts of Eastern Europe (Basabe and Ros 2005). It should be noted that some individualistically oriented people live within collectivistic nations and vice versa, that the individualistic–collectivistic spectrum is a shorthand or framework but not meant to define all people living within any nation.

The individualistic–collectivistic spectrum offers one explanation for why it is easier to gain support for public works in some locations, why in other places there is a greater concern for private property and privacy, why eminent domain is frequently contentious, and why town planners, engineers, and project leads may have conflict with people who seemingly fail to understand the public good being offered – or perhaps more accurately fail to value the public good if it comes at a significant personal expense. An understanding of a culture's perspective, as well as the perspective of the individual person or organisation, can help with framing a potential project in a way that appeals to that perspective's sensibilities. Knowing whether to appeal to the benefits to the individual or the collective good can make the difference in whether a project meets resistance or gains support.

REFERENCES

Abdollahi, M. 2021. "Economic Sanctions and the Effectiveness of the Global Climate Change Regime: Lessons from Iran." *Climate Change Law and Policy in the Middle East and North Africa Region*, 119–135. London: Routledge.

Basabe, N., and M. Ros. 2005. "Cultural Dimensions and Social Behavior Correlates: Individualism-collectivism and Power Distance." *International Review of Social Psychology* 18(1): 189–225.

Kenfack, C. E. 2022. "The Paris Agreement Revisited: Diplomatic Triumphalism or Denial of Climate Justice?" *Journal of Environmental Protection* 13(2): 183–203.

Lawrence, M. G., S. Schäfer, H. Muri, V. Scott, A. Oschlies, N. E. Vaughan, O. Boucher, H. Schmidt, J. Haywood, and J. Scheffran. 2018. "Evaluating Climate Geoengineering Proposals in the Context of the Paris Agreement Temperature Goals." *Nature Communications* 9(1).

Morton, S., D. Pencheon, and N. Squires. 2017. "Sustainable Development Goals (SDGs), and Their Implementation National Global Framework for Health, Development and Equity Needs a Systems Approach at Every Level." *British Medical Bulletin*: 1–10.

Nartova-Bochaver, S. K., M. Donat, G. K. Ucar, A. A. Korneev, M. E. Heidmets, S. Kamble, N. Khachatryan, I. V. Kryazh, P. Larionow, and D. Rodríguez-González. 2022. "The Role of Environmental Identity and Individualism/collectivism in Predicting Climate Change Denial: Evidence from Nine Countries." *Journal of Environmental Psychology* 84: 101899.

Pauw, W. P., P. Castro, J. Pickering, and S. Bhasin. 2020. "Conditional Nationally Determined Contributions in the Paris Agreement: Foothold for Equity or Achilles Heel?" *Climate Policy* 20(4): 468–484.

Stewart, M. G. 2023. "Climate Adaptation Engineering: An Optimist's View." *ASCE-ASME Journal of Risk and Uncertainty in Engineering Systems, Part A: Civil Engineering* 9(1): 02522002.

Triandis, H. C. 2001. "Individualism-Collectivism and Personality." *Journal of Personality* 69(6): 907–924.

Xiang, P., H. Zhang, L. Geng, K. Zhou, and Y. Wu. 2019. "Individualist – Collectivist Differences in Climate Change Inaction: The Role of Perceived Intractability." *Frontiers in Psychology* 10: 187.

Sociological Changes 4

Demographic Trends in the Ways That People Live, Work, and Behave

---

4.1 INTERPRETING DEMOGRAPHIC TRENDS

Demographic trends, at their core, are about births and deaths, about whether a community (or nation, or the world) is experiencing population growth or decline, and the ways in which that population is changing with regard to other characteristics and life events such as marriage. However, there are many other factors at play, including population-limiting issues like disease, famine, and war; purposeful population control through reproductive health measures; and healthcare and healthy lifestyles leading to greater longevity.

The work of demographers often involves tracking cohorts of people – those born in the same place at the same time – to follow them throughout their lives to understand rates of cohabitation, marriage, divorce, disease, and migration, amongst other issues. Demographers also study and interpret data on *who* is reproducing and at what rate; in countries where women are more

                    DOI: 10.1201/9781003392194-4

highly educated, birth rates are lower because women spend a longer period of time in formal education, delaying marriage and tending to be older when they marry and begin having children. Overall, there has been a decline in the rate of teenage pregnancies. Whilst more education, an increase in participation in the work force, and advancements in reproductive technologies have meant that more women are giving birth at 40 years and older than in prior generations, this increase, however, does not offset the decline in teenage pregnancies (Smock and Schwartz 2020). Demographers study these kinds of changes to make forecasts about, for example, whether a population will increase or decrease in size. In addition to age and fertility, demographers interpret data on race, ethnicity, religion, migration (both immigration and emigration), educational status, employment, education, and other characteristics of people.

The point of demographic research is largely to understand how populations of people are changing and what implications those changes have for communities. Demographers pose questions such as, what is the housing and home ownership status of millennials? With that particular question in mind, they would seek to understand the average age of millennials when they purchase their first homes; the size, type, and value of the dwelling in which they live; what percentage of millennials cohabitate with romantic partners, with roommates, with parents, or live alone; whether they live in urban centres, in the suburbs, in towns or villages, or in rural areas, and whether these locations reflect their expressed preferences, the housing supply, affordability, or something else.

A population, in statistical terms, means all members of a group. For a city, the population would be all residents of that city, regardless of their legal status or whether they are homeowners, renters, lodgers, or unhoused. For a primary school, the population of pupils would be all students currently enrolled at that school; this population could be divided into cohorts, most easily as each year's class of pupils. Many organisations, as varied as schools and marketing companies, study their own population or target audience, in efforts to understand the present population and its needs and to forecast future changes and predict changing needs. To return to the example of the trend of women becoming mothers at older ages, the use of reproductive technology has also meant an increase in the frequency of births of twins or other multiples (Fell and Joseph 2012). A company that is manufacturing baby equipment, such as baby strollers/buggies, would investigate their market audience to find that products for twins and triplets would be increasingly desired. Demographers supply the data and interpretation of that data that aid in making decisions whether it is for a company or a city.

For many people, the most familiar source of demographic data come from their country's national census. A census is different from a survey (which, again, could be administered by a company, school, city, or other organisation),

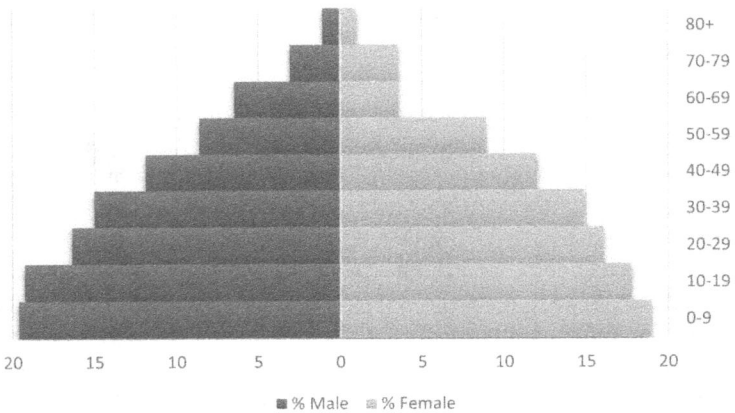

FIGURE 4.1   Population Pyramid of a Country with a Large Percentage of Young People and a Small Percentage of Elderly People.

because the latter involves a sample – the data are drawn only from some percentage of people (or households, or some other unit of analysis) that make up the population. A census, however, is intended to be a complete count of all members of a population, at the same time, and occurring at predictable intervals (e.g., every ten years) to allow for comparisons of sets of data (Baffour et al. 2013). Different countries have different methods of conducting their censuses, including how frequently the data are updated, but the goal remains the same – to have a comprehensive registry of the country's population.

Within a geographic area, such as a nation, city, or neighbourhood, demographers depict the population using a population pyramid, also called an age–sex pyramid, which divides a given population into two halves (male and female) and amongst age cohorts (e.g., 0–9 years, 10–19 years, and so on). In a place where there are many young children and proportionately few elderly people, the population pyramid will look something like a triangle, with the peak at the top, such as in Figure 4.1. A country such as Ghana fits these characteristics because of the high proportion of children and younger adults and low proportion of older adults (GSS 2010).

4.2 FORECASTING FUTURE CHANGES

Having a reliable forecast of how a population will change is essential knowledge in planning services and infrastructure to serve that future population. After all, construction works often take years to complete; a bridge is not built to only meet the needs of today's users but to meet the demands of users in 50 years or more. Regardless of whether the geographic unit is a country, county, city, or neighbourhood, the equation for population change remains the same: Future population equals the baseline population plus births, minus deaths, plus immigration, minus emigration (Wolf and Amirkhanyan 2010). Engineers are unlikely to be asked to compute an area's population or to forecast population changes, but nevertheless it may be useful to understand how the figures are calculated.

For the most part, changes happen predictably and smoothly. Under ordinary circumstances, outside of wartime, the population most at risk of dying is the elderly. In any given place, the population of the elderly will already be known as they are followed through censuses from the time of their birth. Similarly, in any given time interval, such as 2020–2030, the population of women who could give birth is already established, as these individuals capable of giving birth are already alive (Wolf and Amirkhanyan 2010). Changes in rates of reproduction, as referenced earlier as a result of changes in rate of education and women's participation in the work force, tend to take place gradually.

Demographers will forecast population growth if changes in the aforementioned population balancing equation mean that there is a net increase. However, this needs to be further dissected. The growth may be because of advances in healthcare, leading to greater longevity, and fewer elderly people dying in any given time period, even though death is inevitable. For example, a country such as Japan currently has a top-heavy population pyramid where there are proportionately many elderly people and proportionately few children. Life expectancy is amongst the highest in the world, which explains why the elderly population is so large, but the reason for the population is top heavy is because there is a low level of fertility, with fertility rates having dropped off after World War II (Muramatsu and Akiyama 2011). When there are low levels of fertility, smaller birth cohorts result. With each smaller cohort, the subsequent cohorts tend also to be small, as there will be fewer people who are reproducing.

The replacement rate of fertility, or RRF, for a population is defined as how many children, on average, a woman needs to have for the population to remain stable, and this rate is often used interchangeably with the total fertility rate, or TFR. This rate is traditionally set at 2.1, as women constitute approximately half the population and not all infants survive to live to an age to reproduce (Gietel-Basten and Scherbov 2020). This TFR of 2.1 is used as a shorthand, with the acknowledgment that exact sex ratios and mortality rates would be needed to know the true RRF in a given population, as it could be greater or lesser than 2.1. In countries with high infant mortality rates, for example, a TFR of much higher than 2.1 is needed for it to equal the RRF.

With smaller birth cohorts such as in Japan, demographers are able to forecast a top-heavy shape that will continue unless younger cohorts are convinced to have more than 2.1 births per woman, immigration is increased by a large enough margin to offset a low birth rate, or emigration is decreased by a large enough margin. However, given the severe top heavy shape to Japan's population pyramid, a decrease in emigration would be insufficient to appreciably alter the shape of the population pyramid; this change would need to be in combination with one or both of the other two changes mentioned. The consequences of top-heavy populations will be discussed in the next section.

Certain events, such as war, famine, natural disaster, and disease, can make a significant impact on a population. War can result in increased morbidity, mortality, and disability; an increase in migration both within a nation or region and to other nations as refugees; and lowered fertility and marriage rates (Mizoguchi 2010). Natural disasters are often linked to famine and disease, which lead to greater morbidity and mortality for all members of a population but especially the older cohorts as well as infants and children.

4.3 UNDERSTANDING THE IMPACT OF SOCIOLOGICAL CHANGES

Demographic trends are more than a mere academic study in which changes are recorded. Changes in a population can have meaningful and long-lasting impacts in the way that people live, work, and recreate. For example, in the United States during the post-World War II era, there was a significant shift in the general population away from urban areas and into newly built suburban locations, which was made possible by the building of extensive motorways (highways) and the federal government's backing of home mortgages, which were used for securing these suburban homes (Weisbrod et al. 1980). This shift coincided with the Baby Boom, which was a surge in the birth rate, but the popularity of the suburbs remained throughout the 20th century. Researchers have posed the question of whether any subsequent generation would ever return to favouring city life over suburban life, or whether the preference for a suburban lifestyle, which generally includes a larger living space and private garden, reflected the preferences of specific generations or the preferences of each generation upon reaching a certain age, marrying or cohabitating, and having children (Lee 2022).

Deciphering whether a shift in the population to away from urban cores and into the suburbs is likely to be temporary or permanent is necessary to plan for appropriate infrastructure. Where people live, work, and recreate, or where they are forecast to engage in these activities in the future, is likely to be where infrastructure needs to be reinforced, and along the paths between those locations is where public transit will need to be enhanced and where roads and bridges may need to accommodate greater traffic. Conversely, in areas experiencing population decline, so-called shrinking cities, where the population is not forecasted to rebound to peak levels for the foreseeable future, there may be a need for strategic divestment in the area since as the population decreases, property tax revenue to fund public works may also decrease (Morckel and Rybarczyk 2018).

The City of Flint, Michigan, in the United States, serves as a perfect example of a shrinking city. Once home to a thriving economy based on automobile production, key manufacturers began leaving the area in the 1970s, leading to a loss of employment opportunities (Hollander 2010). Urban renewal practices during this time began to displace residents of Flint, disproportionately affecting Black residents, leading to disruption of neighbourhood communities and a devaluation of properties (Highsmith 2009). With weak regional planning practices in the United States, there were no sufficient plans to address shifts in the economic base and population of Flint (Morckel 2017). The case of Flint is unusual because of the health consequences that resulted from a series of poor judgments from government officials. In an effort to save money, the city switched its water source from a less acidic source, Lake Huron, to a more acidic source, the Flint River, but failed to add corrosion control chemicals to the water (Torrice 2016). This failure caused lead and other heavy metals to be leeched from the city's pipes into the drinking water supply, resulting in long-term health consequences, such as lead poisoning, amongst the city residents, especially children (Morckel 2017, Morckel and Rybarczyk 2018). Although Flint has made the international news because of the severe consequences of their water crisis, many European and North American cities have aging water infrastructure where decisions will need to made about how and when to repair or replace it (OECD 2015).

Additionally, the flight from urban cores is not universal; an understanding of the specific context in which one is working is key. For example, in the post-World War II era in Japan, young adults, newly married and starting families, moved from rural areas to urban centres, especially Tokyo. Over decades, these families have remained in metropolitan areas, and thus many elderly Japanese now reside in urban communities (Muramatsu and Akiyama 2011). Japan is challenged by both an aging and shrinking population, as well as one that is increasingly urbanised, leading not only to a predicted shortage of workers and consequences for the nation's economy, but also to a loss of rural communities (Dilley et al. 2022). Rural life is seen as unambitious, but the Japanese government is seeking to change the way that rural life is perceived and to encourage migration out of urban centres and into the countryside – specifically amongst the younger generations and remote workers (Reiher 2020).

4.4 LINKS TO THE UNITED NATIONS' SUSTAINABLE DEVELOPMENT GOALS

As previously stated, when the length of time that girls and women spend in education increases, there are demographic effects – namely, the average age at first marriage increases and the TFR decreases. The Sustainable Development Goals (SDG) set targets that concern these very issues, including reproductive health and female education. For example, SDG4 includes ten targets, such as 4.1, which is for all girls and boys to complete high-quality secondary education, and 4.6, for a substantial portion of all adults to be literate and numerate. With regard to fertility, target 3.7 seeks to ensure 'universal access to sexual and reproductive health-care services, including for family planning, information and education and the integration of reproductive health into national strategies and programmes.' If these goals are met, the rate of population growth should decline (Abel et al. 2016).

Aside from education and fertility, migration can have a significant effect on the population of regions and nations if they experience high rates of immigration or emigration, although there is no effect on the total global population. Migration can be a politically loaded topic in any number of countries, including in the United States, where the Baby Boom was the end of population growth from high numbers of births; instead, population growth is largely spurred from immigration. Political extremes on both sides, however, argue for limits on immigration, albeit for different reasons (Kolankiewicz et al. 2015). In the United Kingdom, for example, the UK Independence Party ran an anti-immigration campaign, leading to the result of the referendum to leave the European Union (so-called 'Brexit') based on an argument that immigration was a threat to the British people and economy (Cap 2019).

Regardless of political persuasion, immigration poses challenges to meeting the SDGs. Whilst highly educated, highly skilled workers may aid their new countries in achieving their SDGs, refugees and asylum seekers arrive in their new countries in need of shelter, education, and other provisions, making it harder to meet various targets (Helgason 2020). Nevertheless, they may also help with other problems that the host countries face; they may help lower the average age in an aging society, posing less of a burden on healthcare systems that are challenged by aging populations; they can contribute their labour, especially in areas where there are critical shortages; and they pay taxes, contributing to social security funds that may be depleted by an aging population. Additionally, the spirit behind the 2030 Agenda was to improve life for all people everywhere, or 'Leave no one behind,' which must be interpreted to include immigrants, whether voluntary or involuntary (Denaro and Giuffré 2022).

4.5 HOW ENGINEERING HAS ENABLED DEMOGRAPHIC CHANGES

Engineering represents the world's great hope in achieving sustainability, by finding solutions to humanity's needs for adequate transportation, sanitation, water supplies, energy sources, and more. However, engineering has also fuelled many of the changes that society has experienced since industrialisation and urbanisation began, through the building of better transportation systems, communication systems, and other infrastructure that has simultaneously allowed for greater density and greater dispersion (Marjoram 2019). We have engineers to thank and blame for the invention of the automobile, without which population development patterns would be quite different, as there would be no high-speed motorways, urban sprawl, large-scale housing developments, suburban or exurban communities (besides those reachable via public transport), car parks and garages (taking up valuable land and posing challenges to density), and so on.

In addition, through advancements in technology that have allowed for cleaner drinking water, better sanitation, and widespread electricity, populations have been kept healthier and warmer, aiding in greater longevity and a better quality of life. For better or worse, without innovations in engineering, we would not be able to maintain anywhere close to our current levels of human population. This, at its root, is the Malthusian debate – to what extent can engineering, science, and technology continue to develop solutions to support not only an increasing population but also a population that consumes resources at increasingly higher rates?

REFERENCES

Abel, G. J., B. Barakat, S. K., and W. Lutz. 2016. "Meeting the Sustainable Development Goals Leads to Lower World Population Growth." *Proceedings of the National Academy of Sciences* 113(50): 14294–14299.

Baffour, B., T. King, and P. Valente. 2013. "The Modern Census: Evolution, Examples and Evaluation." *International Statistical Review* 81(3): 407–425.

Cap, P. 2019. " 'Britain is Full to Bursting Point!': Immigration Themes in the Brexit Discourse of the UK Independence Party." In *Discourses of Brexit*, 69–85. London: Routledge.

Denaro, C., and M. Giuffré. 2022. "UN Sustainable Development Goals and the "Refugee Gap": Leaving Refugees Behind?" *Refugee Survey Quarterly* 41(1): 79–107.

Dilley, L., M. Gkartzios, and T. Odagiri. 2022. "Developing Counterurbanisation: Making Sense of Rural Mobility and Governance in Japan." *Habitat International* 125: 102595.

Fell, D. B., and K. Joseph. 2012. "Temporal Trends in the Frequency of Twins and Higher-order Multiple Births in Canada and the United States." *BMC Pregnancy and Childbirth* 12: 1–7.

Gietel-Basten, S., and S. Scherbov. 2020. "Exploring the 'True Value'of Replacement rate Fertility." *Population Research and Policy Review* 39(4): 763–772.

GSS, G. 2010. "Population and Housing Census: Summary Report of Final Results." In *Accra: Ghana Statistical Service*. Accra, Ghana: Ghana Statistical Service.

Helgason, K. S. 2020. *The Economic and Political Costs of Population Displacement and Their Impact on the SDGs and Multilateralism*. United Nations, Department of Economic and Social Affairs. New York.

Highsmith, A. R. 2009. "Demolition Means Progress: Urban Renewal, Local Politics, and State-sanctioned Ghetto Formation in Flint, Michigan." *Journal of Urban History* 35(3): 348–368.

Hollander, J. B. 2010. "Moving Toward a Shrinking Cities Metric: Analyzing Land use Changes Associated with Depopulation in Flint, Michigan." *Cityscape*: 133–151.

Kolankiewicz, L., B. Griffith, S. A. Camarota, and R. Beck. 2015. "Immigration, Population Growth, and the Environment." *Center for Immigration Studies*.

Lee, H. 2022. "Are Millennials Leaving Town? Reconciling Peak Millennials and Youthification Hypotheses." *International Journal of Urban Sciences* 26(1): 68–86.

Marjoram, T. 2019. *Engineering for Humanity, Sustainability and the SDGs*. Melbourne: Engineers Australia.

Mizoguchi, N. 2010. *The Consequences of the Vietnam War on the Vietnamese Population*. Berkeley: University of California.

Morckel, V. 2017. "Why the Flint, Michigan, USA Water Crisis is an Urban Planning Failure." *Cities* 62: 23–27.

Morckel, V., and G. Rybarczyk. 2018. "The Effects of the Water Crisis on Population Dynamics in the City of Flint, Michigan." *Cities & Health* 2(1): 69–81.

Muramatsu, N., and H. Akiyama. 2011. "Japan: Super-aging Society Preparing for the Future." *The Gerontologist* 51(4): 425–432.

OECD, W. 2015. *Cities: Ensuring Sustainable Futures, OECD Studies on Water.* Paris: OECD.

Reiher, C. 2020. "Embracing the Periphery: Urbanites' Motivations for Relocating to Rural Japan." In *Japan's New Ruralities*, 230–244. London: Routledge.

Smock, P. J., and C. R. Schwartz. 2020. "The Demography of Families: A Review of Patterns and Change." *Journal of Marriage and Family* 82(1): 9–34.

Torrice, M. 2016. "How Lead Ended up in Flint's Tap Water." *Chemical & Engineering News* 94(7): 26–29.

Weisbrod, G., S. R. Lerman, and M. Ben-Akiva. 1980. "Tradeoffs in Residential Location Decisions: Transportation Versus Other Factors." *Transport Policy and Decision Making* 1(1): 13–26.

Wolf, D. A., and A. A. Amirkhanyan. 2010. "Demographic change and its public sector consequences." *Public Administration Review* 70: s12–s23.

Environmental Matters 5

Protectionism and Protests

5.1 ENVIRONMENTAL IMPACTS OF CIVIL INFRASTRUCTURE

The building industry is responsible for about a third of all carbon emissions, between operational emissions, referring to those emissions that occur when the buildings are being used, and embodied carbon, referring to the carbon emitted in extracting, manufacturing, and transporting materials for buildings to be constructed as well as the carbon emitting during the construction processes themselves (Rodrigo et al. 2019). With the decision that a building or structure is needed comes a multitude of other decisions, including the materials out of which it should be built, how large of a structure it will be and thus how much of the materials it will need to use, and to what extent the building will be able to deconstructed at some later point and its materials reused.

Civil infrastructure uses a high volume of concrete because of its durability, versatility, and strength, yet concrete poses significant challenges to sustainability (Guo et al. 2020). Life cycle assessment (LCA) is a standardised method of calculating the environmental impact of buildings. For multistorey residential buildings and for industrial buildings, LCA has been used to find that timber is a more sustainable material than steel and reinforced concrete (Hegeir et al. 2022). Innovations in materials such as cross laminated timber

    DOI: 10.1201/9781003392194-5

(CLT) as flooring are being found to lower levels of carbon emissions and a greater capacity to store carbon when compared with concrete, but have some limitations, such meeting structural requirements at greater spans (Hassan et al. 2019). Research on materials continues to investigate ways to make concrete more sustainable and to make use of other materials, such as waste glass (Kazmi et al. 2020).

In addition to the choice of materials, an always important consideration is the location. Attention must be given to what was previously on the site, what is currently on the site, what is nearby, and what is in the path between the site and the origins for likely users of the site. For certain projects, such as roads, the path that they will take may be extensive and cutting through various ecosystems.

For example, lakes and rivers provide valuable ecological services to a community, yet also sometimes pose a challenge for transporting goods and people across. A bridge or tunnel is built for the purpose of improving transportation, but a proposal to build a bridge or tunnel must consider more than the change in vehicular traffic but also environmental risks and impacts during the construction phase such as habitat destruction, loss of wildlife species, increased soil erosion, water pollution for the body of water as well as groundwater, and induced traffic demand leading to increased suspended solids (Li et al. 2022). Increased water pollution from petrochemicals and heavy metals in surface runoff may decrease the water quality and require further treatment, at an expense, before that water can be used by the community (e.g., as a potable water source).

These choices that civil engineers make in designing and maintaining structures relate to the four Sustainable Development Goals that are the most focused on the natural environment: SDG 12 Responsible Consumption and Production, SDG 13 Climate Action, SDG 14 Life Below Water, and SDG 15 Life on Land. SDG 12 highlights the importance of everything from LCA to developing innovative uses for waste materials to having cleaner transportation technologies. Target 12.2, for example, concerns the sustainable management and use of natural resources (e.g., timber). SDG 13 is the broadest of all the SDGs with regard to climate change, setting targets for combatting extreme weather conditions, limiting greenhouse gas emissions, and using renewable energy. Target 13.1, 'Strengthen resilience and adaptive capacity to climate-related hazards and natural disasters,' is measured through the passing of national and local policies and plans in which engineers may be involved in developing or with which they will need to comply. With regard to both SDG 14 and 15, everything built will either affect the quality of the oceans and other waters and/or the quality of land, either directly or indirectly, whether it is a building, a bridge, or a dam. Civil engineers have a responsibility to protect the biodiversity of ocean-dwelling and land-dwelling organisms through responsible and sustainable construction.

5.2 NET ZERO, CARBON NEUTRAL, AND CARBON NEGATIVE

Several terms compete for attention within the space of environmental considerations of the built environment. These terms are often used interchangeably or in a vague way that alludes to more environmentally friendly practices but without specific definitions. Perhaps most confusingly, some sources will use these terms incorrectly, exchanging one term for another and swapping the definitions.

Three related terms – net zero, carbon neutral, and carbon negative – by definition rely on the same scientific premise: that greenhouse gas emissions, especially carbon dioxide, lead to an increase in the amount of greenhouse gases in the planet's atmosphere, which in turn leads to a rise in the planet's temperature. The temperature of the planet is on a trajectory to rise to dangerous levels that will make the planet unhospitable for sustaining human life, as well as the lives of other organisms. The terms net zero, carbon neutral, and carbon negative are used in everything from policy documents to descriptors for buildings, but no matter the context, the terms reflect a concern for the climate crisis and a goal or target to limit the warming of the planet.

For an organisation (whether it is a city or a for-profit company) to be carbon neutral, it needs to cancel out its carbon emissions, generally through the purchasing of offsets, or carbon credits. This means that there is a reduction or removal of carbon (and other greenhouse gases) elsewhere. Carbon offsetting is an exceedingly contested approach because it does not require any commitment to change or reduce the organisation's GHG emissions; the organisation can carry on its everyday practices as normal. To use a simplified example, it can be thought of as acknowledging the carbon emissions resulting from a company's air travel, but rather than taking fewer airplane trips, a number of trees are planted annually to serve as an equivalent carbon sink – as compensation for each year's air travel. Net zero, on the other hand, takes a different approach because net zero means that the organisation first resolves to reduce their emissions and then offsets any remaining emissions. Net zero approaches, therefore, are the better solution for the environment (Herbert Smith Freehills 2020).

Some organisations take it a step further and pledge to be carbon negative in their practices. This means that an organisation removes more carbon from the atmosphere than it emits. For example, Microsoft announced plans to be carbon negative by 2030 and, by 2050, to have removed additional carbon from the atmosphere equivalent to all carbon the company has emitted since its founding in 1975 (Budinis 2020). At present, pledges such as Microsoft's are still outliers, as the most common pro-environment goal is to be carbon neutral.

5.3 THE CASE FOR REUSE, REBUILDING, AND BROWNFIELD SITES

Undeveloped land generally represents the most straightforward, uncomplicated land on which to build, in contrast with sites that have been previously developed, especially when they had been used for commercial or industrial uses and then abandoned, resulting in 'brownfield' sites (De Sousa 2000). The risk of sites such as these ones being contaminated, necessitating remediation that could be costly and time consuming, or possibly the site even being unusable for redevelopment, can lead potential developers to shy away from choosing these sites in favour of the undeveloped, greenfield sites. Even though brownfield sites are not necessarily contaminated, there is often a perception that they are and often that perception is enough to dissuade potential developers from redeveloping those sites (De Sousa 2000).

The reasons for choosing brownfield sites, or any other sites where a retrofit, conversion, or reuse/rebuild on the existing site would be possible, are plentiful and are based on pragmatic as well as philosophical reasons. Although all environmental, economic, and social costs and benefits should be considered before choosing a previously developed site, the first and foremost reason to choose such a site is that it is likely to already be connected to infrastructure (Williams and Dair 2007). Beneath an urban industrial site, for example, there may be water and gas pipelines, electricity cables, and sewer lines that were previously built to service the site and now lie used; these utilities are further connected outside of the site to transformers, pumping stations, power stations, and so on (Moss 2003). In addition, the site will already have some access to transportation, whether that includes road, rail, canals, or a port. This connectedness to transportation is of course essential for the production and shipping of goods, but is equally important if the site concerns, for example, a warehouse conversion into residential spaces, where the new inhabitants will need to be able to walk, bike, drive, or take public transport to job centres, recreational areas, shopping, etc.

In line with the policies of many developed nations around the world, the redevelopment of previously developed sites in urban locations is key in protecting rural areas as well as promoting density and regeneration where it is most appropriate – in the cities. Encouraging the (re)development of urban sites is a way of fostering compact city form, preventing low-density sprawl into greenfield areas (Talen 2011). Rather than leaving derelict or underused structures in place, the opportunity exists to rebuild or retrofit these spaces to add density, activity, and vibrancy to the web of existing places.

In developing in such a manner, greenfield sites remain untouched, and environmental stewardship can be prioritised. Preserving greenfield sites protects the biodiversity of those sites, protects areas that are hydrologically sensitive, maintains wildlife corridors, and conserves areas of natural beauty both for the sake of the conservation itself (since development of untouched land is an irreversible action) and for the enjoyment and recreation of people. Hosts of locations as diverse as Copenhagen, Denmark, to Melbourne, Australia, to Reston, Virginia (United States), have proactively taken measures to protect greenfield and greenbelt areas, recognising the importance of preserving natural spaces and preventing the encroachment of the urban areas (Buxton and Goodman 2003).

To rebuild, reuse, and redevelop on existing sites, and in doing so curbing sprawl, has another environmental benefit: preventing further automobile dependency. The environmental benefit comes out of a cycle of how typical sprawl development, being low density and on the urban fringe by definition, generally requires the use of cars, which leads to transport infrastructure being designed to accommodate more cars. This means wider streets, less walkability, and less use for public transport – which may be offered less frequently and at a higher cost to account for it being less widely used. This feeds back into a greater reliance on cars, which leads to pollution, smog, and congestion in cities, which then feeds into a greater pressure to sprawl and develop into less developed areas. And so the cycle continues.

5.4 REGENERATIVE DESIGN AND DESIGNING WITH NATURE

Most readers will be familiar with terms such as green buildings and sustainable design. Architects and related professionals will label a building as 'green' if it performs better on any variety of environmental metrics or indicators, as compared to a conventional or typical building, especially with regard to using renewable energy, building with materials that have a lower environmental impact, and having more efficient heating and cooling systems. In general, a green building is supposed to cause less harm to the environment than an equivalent traditionally designed building (Cole 2012).

The terms sustainable and sustainable development can also be applied to green buildings, although the terms can also be used to more explicitly reference the Brundtland Report or more generally to imply a building or structure that is better for the environment than a traditional design. 'Sustainable,' even more than 'sustainable development,' is a term that the general public understands. However, there is also ambiguity resulting from the term's widespread use, and there is also criticism coming from academics who posit that to be sustainable is not enough and that sustainable development is an oxymoron – that the pursuit of economic goals hinders environmental goals and cannot be reconciled (Spaiser et al. 2017). This criticism springs from decades-old arguments from ecological/environmental economists who advocate instead for a steady-state economy (Daly 1977).

One perspective is that the term 'sustainability' can and should be evolved to be more wholistic and integrative, where it can address economic, environmental, social and other (e.g., political/power) concerns (Robinson 2004). An opposing perspective is that 'sustainability' and especially 'sustainable design' should be replaced by 'regenerative design.' Proponents of regenerative design view sustainable design as a neutral option – better than green and certainly better than conventional designs but still not going far enough to create environmentally responsible designs (Reed 2007). Their emphasis is on connection to place and specifically the natural environment, with people as participants in the evolution of living systems – in other words, designing with nature. Proponents ardently argue for what they view as a much-needed transformative approach (Lyle 1996, Gibbons 2020, Camrass 2022). The goal of regenerative design is to create healthy relationships between social and ecological systems, going beyond the conventional definition of sustainable design, which proponents of regenerative design see as focusing on minimising negative effects, to actively working to improve natural resources and systems.

In regenerative designs, the complete lifecycle of the project is taken into account, renewable energy must be used, biodiversity must be supported or enhanced, water conservation must be promoted, and local communities must be included and empowered.

One point of view is that it is important to distinguish between 'regenerative' and 'sustainable' because they refer to distinct strategies and objectives for tackling environmental and social problems. The term 'sustainability' refers to actions that minimise harmful effects on the environment, the economy, and society whilst aiming to maintain or improve the well-being of current and future generations. It emphasises preserving resources, reducing waste, and maintaining long-term survival. 'Regenerative' connotes actively restoring and regenerating existing systems and stresses the need to take a proactive and transformative approach. Proponents of regenerative design may view sustainability as passive.

Although there are differences between regenerative and sustainable techniques, their ultimate objectives frequently overlap. Both perspectives aim to address societal and environmental problems, encourage smart resource management, and protect the welfare of the present and future generations. Regardless of the terminology used, efforts should be focused on taking practical steps that enhance environmental and social results rather than debating semantic differences or introducing additional complexity and confusion in terminology that could hinder progress in policy-making.

Terms evolve and it might be true that 'regenerative design' will come to replace 'sustainable design.' However, all terms and concepts are subject to criticism, and even 'regenerative design' has received some criticism for being perceived to have reinvented the wheel – that the term has a similar meaning to other concepts, those of cradle-to-cradle design and biophilic design (Andreucci et al. 2021). A separate criticism of the regenerative design concept is that regenerative design, along with permaculture, biomimicry, and biophilic design are insufficient for preserving nature; they represent a closed system in which nature co-evolves with humans, which critics argue is impossible – as evidenced by the loss of biodiversity and other markers of nature's failure to keep pace with the impacts of the climate crisis (Birkeland 2020). Finally, whilst restoring the local ecology of an area and leaving the site better than it was found are worthy and admirable goals, the concept of regenerative design must be widened to show its relevance to social/environmental justice. Some proponents of regenerative design have begun to expand the term to encompass these social issues.

5.5 LEED AND BREEAM

In recognition of the environmental impact that buildings have, a number of building certification systems have been created to assign a rating to each building or design based on its energy efficiency and environmental impact. Most famously, LEED, or Leadership in Energy and Environmental Design, was created by the U.S. Green Building Council, a non-profit organisation. Originally formed as a green building certificate system specific to new construction, over the years it has been expanded to the following categories: new construction or major renovations (Building Design and Construction), interior design (Interior Design and Construction), existing buildings including schools, retail, hospital, warehouses (Building Operations and Maintenance), new or redeveloped neighbourhoods of mixed use or residential use (Neighborhood Development), residential buildings including single-family residences as well as multifamily (Homes), and city systems such as waste and transportation (Cities) (Amiri et al. 2019). Within the LEED system, four levels of certification are possible – Certified, Silver, Gold, and Platinum – with each higher level associated with more points earned on items associated with, for example, water efficiency or the materials used in construction.

LEED is one of the most widely used green building certification systems around the globe, in part because of its length of history but also because of its applicability to nearly every type of construction and design retrofit (Amiri et al. 2019). It has been criticised, however, for being expensive, with fees associated with the application and review of a submission, as well as for being unnecessarily complicated and time consuming (Leontev 2021). An additional criticism of LEED, although one that could extend to any similar rating system as well, is that the public relations incentive for having a 'green building' can lead to situations like moving a company's headquarters out of an existing building in a dense, urban location and into a new, LEED-certified building in a greenfield, suburban location – so that although the new building has a lower environmental impact than the former building, the overall, net impact on the environment is greater.

The other major certification system to know, especially if working in the UK, is BREEAM (Building Research Establishment Environmental Assessment Method). Predating LEED, which arguably has its emphasis on energy efficiency, BREEAM focusses more on environmental sustainability as a whole. Like LEED, BREEAM has certification categories for different kinds of projects, including: BREEAM New Construction (for new non-residential buildings in the UK), BREEAM International New Construction (for new residential and new non-residential buildings in countries other than the UK or where there is a country-specific BREEAM scheme), BREEAM In-Use (for running costs and environmental performance of existing buildings), BREEAM Refurbishment (for residential retrofits), BREEAM Communities (for master planning), and Home Quality Mark (for residential buildings in the UK).

BREEAM and LEED differ in several ways. LEED certification takes place after a project is built, whereas BREEAM certification occurs at two points: in the design phase as well as once the project is completed. Those seeking LEED certification are responsible for preparing and filing the application and evidence themselves, although they can hire consultants who specialise in LEED. BREEAM certification involves an external assessor, rather than an internal employee of the engineering/architecture firm.

Perhaps the most importance difference is that LEED's standards are rooted in American codes (from the American National Standards Institute [ANSI] and the American Society of Heating, Refrigerating and Air Conditioning Engineers [ASHRAE]) whereas the reference standards for BREEAM come from the International Standards Organization (ISO) and European Standard Organization (CEN) (Ferreira et al. 2023).

5.6 CEEQUEL AND ENVISION

Whilst LEED and BREEAM have been developed to assess the sustainability of the built environment beyond individual buildings, their primary application is for those individual buildings. Historically, there have been two main rating systems that are analogous to LEED and BREEAM for civil engineering, public realm, and infrastructure projects such as transport, water, and energy: CEEQUAL and Envision, although other rating systems also exist. Since 2015, the BRE Group, which operates BREEAM, absorbed CEEQUAL and rebranded it as BREEAM Infrastructure. Developed in 2003 by the Institute of Civil Engineers (ICE) in the UK, many civil engineers are likely to continue to refer to BREEAM Infrastructure by its former name (Griffiths et al. 2017).

Whether called BREEAM Infrastructure or CEEQUAL, the scheme works much the same way as other rating systems (e.g., LEED), in that a detailed assessment evaluates a range of sustainability issues, such as energy use and efficiency, waste management, and biodiversity, with those projects that achieve a high score being awarded a rating that signifies excellence in sustainable infrastructure. Although the scheme originated in the UK, it has projects in other countries as well, including in the Middle East, in Southeast Asia, elsewhere in Europe, and in the US (Mattinzioli et al. 2020).

Envision is quite similar to CEEQUAL/BREEAM Infrastructure, in that it is a rating system for the same types of projects (such airports, railway infrastructure, energy- and water-related projects) and has projects in various countries. Envision, however, was developed by and is administered through the Institute for Sustainable Infrastructure, in the US; is a more recent scheme, having been developed in 2012; and purportedly places more of an emphasis on a wholistic lifecycle analysis (Mattinzioli et al. 2020). In practical terms, whether one uses CEEQUAL/BREEAM Infrastructure or Envision, the result is a rating to show that the project achieved merit in sustainability as measured by a third-party company.

5.7 RESISTANCE AND OPPOSITION TO CONSTRUCTION

Individuals and stakeholder groups may oppose construction projects for any number of reasons. They may also oppose, and try to block, specific projects because they object to the type of project under all circumstances, such as the building of additional airports or of nuclear power facilities, or they may oppose, and try to block, specific projects because they object to the location, which they may view as being inappropriate. Additionally, as discussed earlier in this book, a NIMBY or related outlook could mean that any construction in a specific location would be opposed by certain individuals or stakeholder groups.

Objections may be raised based on a variety of environmental reasons. Objectors may be concerned about habitat destruction and the loss of biodiversity, especially if a particular ecosystem is seen as fragile or if natural habitats in the region are scarce. This can be related to another objection: the loss of green space or open areas, both for the effect on the ecosystem as well as for the loss of an aesthetic or recreational amenity for nearby residents. Specific to the local area, there may be objections to the levels of pollutants resulting from construction and the extent to which they will degrade the local water and air quality. Both during construction and after the project is completed, there may be an increase in traffic, contributing to congestion and air pollution. Finally, objections may be raised based on higher level concerns, such as the impact on the climate crisis that results from the production of building materials and the operational emissions of the structure. Similarly, objectors may be concerned about resource depletion, as the production of steel and concrete contributes to the depletion of water and minerals.

The above reasons are some of the most common environmental reasons that individuals may argue against a project. In planning any development project, the possible effects should be thoroughly evaluated and any harm to the environment or surrounding populations should be mitigated. Conducting a thorough environmental analysis is often required by authorities (such as a local government agency or planning commission), but even when and where it is not required, it may be beneficial if opposition to the project is likely. Individual members of the public and stakeholder groups may contact the relevant authorities themselves. They may write letters of objections outlining their concerns and send these letters to the engineering firm, architecture firm, local government agency, or even the press. They might start petitions against the project. They are also likely to attend public meetings as an opportunity to have their objections heard. Lastly, they may attempt to take legal action, such as filing an appeal to a decision or a lawsuit, to challenge the project. For all these reasons, it is important to be proactively prepared with knowledge about the potential environmental impacts of any project.

REFERENCES

Amiri, A., J. Ottelin, and J. Sorvari. 2019. "Are LEED-certified Buildings Energy-efficient in Practice?" *Sustainability* 11(6): 1672.

Andreucci, M. B., A. Loder, M. Brown, and J. Brajković. 2021. "Exploring Challenges and Opportunities of Biophilic Urban Design: Evidence from Research and Experimentation." *Sustainability* 13(8): 4323.

Birkeland, J. 2020. *Net-positive Design and Sustainable Urban Development*. London: Routledge.

Budinis, S. 2020. *Going Carbon Negative: What are the Technology Options?* Paris, France: International Energy Agency.

Buxton, M., and R. Goodman. 2003. "Protecting Melbourne's Green Belt." *Urban Policy and Research* 21(2): 205–209.

Camrass, K. 2022. "Urban Regenerative Thinking and Practice: A Systematic Literature Review." *Building Research & Information* 50(3): 339–350.

Cole, R. J. 2012. "Transitioning from Green to Regenerative Design." *Building Research & Information* 40(1): 39–53.

Daly, H. 1977. "Steady State Economy." *San Francisco* 545.

De Sousa, C. 2000. "Brownfield Redevelopment Versus Greenfield Development: A Private Sector Perspective on the Costs and Risks Associated with Brownfield Redevelopment in the Greater Toronto Area." *Journal of Environmental Planning and Management* 43(6): 831–853.

Ferreira, A., M. D. Pinheiro, J. de Brito, and R. Mateus. 2023. "A Critical Analysis of LEED, BREEAM and DGNB as Sustainability Assessment Methods for Retail Buildings." *Journal of Building Engineering*: 105825.

Gibbons, L. V. 2020. "Regenerative – The New Sustainable?" *Sustainability* 12(13): 5483.

Griffiths, K. A., C. Boyle, and T. F. Henning. 2017. "Comparative Assessment of Infrastructure Sustainability Rating Tools." Proceedings of the Transportation Research Board 96th Annual Meeting, Washington, DC, USA.

Guo, P., W. Meng, H. Nassif, H. Gou, and Y. Bao. 2020. "New Perspectives on Recycling Waste Glass in Manufacturing Concrete for Sustainable Civil Infrastructure." *Construction and Building Materials* 257: 119579.

Hassan, O. A., F. Öberg, and E. Gezelius. 2019. "Cross-laminated Timber Flooring and Concrete Slab Flooring: A Comparative Study of Structural Design, Economic and Environmental Consequences." *Journal of Building Engineering* 26: 100881.

Hegeir, O. A., T. Kvande, H. Stamatopoulos, and R. A. Bohne. 2022. "Comparative Life Cycle Analysis of Timber, Steel and Reinforced Concrete Portal Frames: A Theoretical Study on a Norwegian Industrial Building." *Buildings* 12(5): 573.

Herbert Smith Freehills. 2020. *Carbon Neutral' and 'Net-zero Carbon': What's the Difference – And Why Does It Matter?* www.herbertsmithfreehills.com/carbon-neutral-and-net-zero-carbon-whats-the-difference-and-why-does-it-matter.

Kazmi, D., D. J. Williams, and M. Serati. 2020. "Waste Glass in Civil Engineering Applications – A Review." *International Journal of Applied Ceramic Technology* 17(2): 529–554.

Leontev, M. 2021. "Analysis of Obstacles to Green Building Projects: The Experience of Russia and Europe." E3S Web of Conferences, EDP Sciences, Les Ulis.

Li, Q., R. Qian, J. Gao, and J. Huang. 2022. "Environmental Impacts and Risks of Bridges and Tunnels Across Lakes: An Overview." *Journal of Environmental Management* 319: 115684.

Lyle, J. T. 1996. *Regenerative Design for Sustainable Development.* Hoboken: John Wiley & Sons.

Mattinzioli, T., M. Sol-Sánchez, G. Martínez, and M. Rubio-Gámez. 2020. "A Critical Review of Roadway Sustainable Rating Systems." *Sustainable Cities and Society* 63: 102447.

Moss, T. 2003. "Utilities, Land-use Change, and Urban Development: Brownfield Sites as 'Cold-spots' of Infrastructure Networks in Berlin." *Environment and Planning A* 35(3): 511–529.

Reed, B. 2007. "Shifting from 'Sustainability' to Regeneration." *Building Research & Information* 35(6): 674–680.

Robinson, J. 2004. "Squaring the Circle? Some thoughts on the Idea of Sustainable Development." *Ecological Economics* 48(4): 369–384.

Rodrigo, M., S. Perera, S. Senaratne, and X. Jin. 2019. "Embodied Carbon Mitigation Strategies in the Construction Industry." CIB World Building Congress, 2019.

Spaiser, V., S. Ranganathan, R. B. Swain, and D. J. Sumpter. 2017. "The Sustainable Development Oxymoron: Quantifying and Modelling the Incompatibility of Sustainable Development Goals." *International Journal of Sustainable Development & World Ecology* 24(6): 457–470.

Talen, E. 2011. "Sprawl Retrofit: Sustainable Urban form in Unsustainable Places." *Environment and Planning B: Planning and Design* 38(6): 952–978.

Williams, K., and C. Dair. 2007. "A Framework for Assessing the Sustainability of Brownfield Developments." *Journal of Environmental Planning and Management* 50(1): 23–40.

Sustainable Economics 6

Is Net Zero Financially Viable?

6.1 UNDERSTANDING THE TRIPLE BOTTOM LINE

To achieve the Sustainable Development Goals (SDGs), a comprehensive strategy that considers economic, social, and environmental issues is necessary. For example, SDG1 aspires to reduce poverty, and achieving this goal will require the creation of good jobs and increased economic opportunities. SDG 8 is focused on advancing full and productive employment, sustainable and inclusive economic growth, and decent work for all. SDG 12 focuses on responsible production and consumption, and is intimately related to economic systems and choices about the use and allocation of resources.

In addition, the SDGs call for the mobilisation of financial and other resources, which is directly related to economic institutions and policies. The private sector must also play a significant role. For sustainable economic growth and development, corporate practices and investment decisions must be in line with the SDGs.

The traditional type of economic theory taught in universities around the world is neoclassical economics, which places a strong emphasis on market-based solutions. For traditional/neoclassical economists, the economy itself is viewed as a self-governing system that responds to market forces and is self-correcting, with minimal need for government intervention or

 DOI: 10.1201/9781003392194-6

regulation, mainly limited to the protection of contracts and property rights. Neoclassical economists base their theories on several assumptions: that individuals act rationally and will seek to maximise their own self-interest, that markets are efficient, and that supply and demand dictates prices. There is also an assumption that resources are limited, but equally a reliance on economic growth to find additional stockpiles of those resources and to discover new resources. For example, neoclassical economists will acknowledge that known coal supplies are limited, but will suggest that the solution is to limited supplies is to mine deeper, open additional mines, and/or invest in other energy supplies (e.g., nuclear power), but not to limit economic growth (Arnsperger and Varoufakis 2006).

As a reaction to neoclassical economics and out of concerns about environmental degradation, environmental economics arose in the 1960s and 1970s (Meadows et al. 2018). Environmental economics can be viewed as falling under the broader category of neoclassical economics, with the additional belief that since there are economics drivers of environmental issues (e.g., degradation, depletion of resources) and thus there must be economics solutions as well (Beder 2011). For environmental economists, the environment is a finite or scarce resource, and neoclassical theories and principles can be applied to evaluate environmental policies.

Ecological economics also emerged in the late 20th century as a reaction to neoclassical economics but, in contrast to environmental economics, is more interdisciplinary, using a systems-based approach to describe the connections between the economy, society, and the environment (Daly 1994). Whereas environmental economists focus on finding market-based solutions to environmental issues, ecological economists see the economy as embedded within the greater biosphere, with economic growth and development affecting the environment, often in a negative way. Where conventional economists value unrestricted economic growth and the extraction of natural resources, ecological economists view this perspective as unsustainable. The goal for ecological economists is to create long-term sustainability through integrating economic, social, and environmental concerns into policy making.

For traditional economists, there is one bottom line – financial performance – for a company or organisation. Whether a company is succeeding is a matter of whether it is operating at a loss or making a profit, and social and environmental issues serve as constraints to making a profit (Anwar and El-Bassiouny 2019). The triple bottom line (TBL), however, evaluates a company's performances in terms of its economic, social, and environmental impacts (see Figure 6.1). A company's economic performance is its financial success, including measures like revenue, profit, and return on investment. Its social performance includes any impact on society, such as labour practices

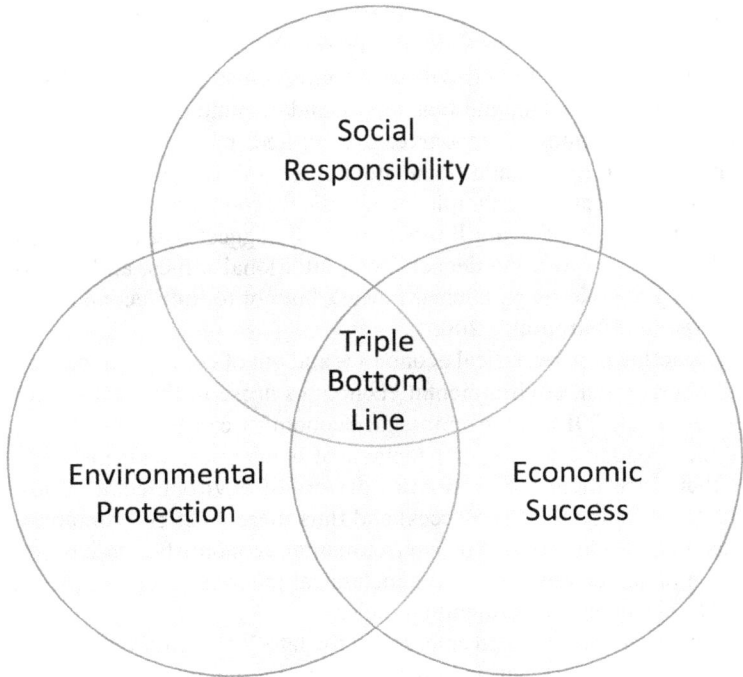

FIGURE 6.1 Triple Bottom Line.

and involvement in the local community. Its environmental performance is based on how the company affects the natural environment, such as its greenhouse gas emissions and resource use.

The Triple Bottom Line approach emphasises that economic, social, and environmental factors need to be taken into account when making organisational decisions. It offers a more comprehensive, integrative picture of an organisation's performance and impacts. This brings the TBL approach in line with ecological economics' objectives, stressing the need to identify long-term policies and solutions that strike a balance amongst economic, social, and environmental issues.

At the other end of the spectrum from traditional economists is an environmentally minded perspective that objects to any emphasis on economic growth, where growing the economy to any extent, or even the need to consider the economy, is anathema to where the real focus should be: on protecting and rehabilitating the natural environment. For those people, sustainable development is an oxymoron. However, in recent years, even this idea has evolved to re-embrace the idea of sustainable development, recognising the value of the widespread acceptance of the term, and focusing on the work of achieving the Sustainable Development Goals (Redclift 2005).

6.2 FOR-PROFIT BUSINESSES

For-profit businesses may find it difficult to balance economic profit and environmental sustainability since short-term financial gains may conflict with long-term environmental objectives. However, increasingly, many businesses are discovering that adopting a strategy that prioritises sustainability may actually help to boost their earnings by cutting expenses, increasing productivity, and attracting customers who are concerned about the climate crisis.

There are several ways for companies to balance economic profit with environmental sustainability, including putting into practice energy-efficient techniques, as well as technology, to save on operating costs. Depending on the nature of the business, it may be possible to find more affordable materials that are also more environmentally friendly and there may be room to cut waste, which would also cut expenses and have a possible influence on the environment. Another strategy lies in the company's marketing: if customers are willing to pay more for environmentally friendly goods and services, a company would do well to advertise their environmentally friendly practices; not only would they be helping the environment but they also have the opportunity to boost their sales.

For-profit companies' environmental concerns and priorities can depend on their industry, location, and size – as well as their overall commitment to being sustainable, since not all CEOs consider environmental sustainability to be important. Companies like Patagonia, The North Face, Google, and IKEA, however, are often cited for their strong environmental policies, whilst continuing to be profitable companies. These businesses have adopted policies like using environmentally materials in their operations and productions, encouraging sustainable practices, and making commitments to lowering their carbon footprint (Xue et al. 2018).

One particularly subcategory of for-profit organisations is worthy of mention: B Corporations. A 'B Corporation,' short for Benefit Corporation, is a type of for-profit business that is legally compelled to take into account how its decisions may affect society and the environment in addition to its shareholders. B Corps must adhere to strict criteria for social and environmental performance, accountability, and transparency in order to receive certification from the non-profit B Lab (Stubbs 2017).

B Corps are different from regular corporations in that they aim to benefit society and the environment whilst also making money for their shareholders. This approach to conducting business seeks to strike a balance between the pursuit of wealth and a dedication to enhancing society by tackling social and environmental problems. The concept behind B Corps is that businesses may be forces for good, and that a corporation's goals should go beyond maximising profits for its shareholders (Honeyman and Jana 2019).

Some examples of B Corps include Allbirds, which is a footwear company that stresses sustainability and transparency in all aspects of its business practices; Warby Parker, which is an eyewear company that operates a 'Buy a Pair, Give a Pair' programme that donates a set of eyeglasses to a person in need with each pair that is purchased; and Dr. Bronner's, a maker of organic personal care products that is renowned for its moral standards, environmental stewardship, and support of social and political causes. B Corps are found across virtually every industry, from beverages (New Belgium Brewing) to books (Better World Books) to social media engagement (Hootsuite) to training and development organisations (The Social Enterprise Academy).

Notably, there are engineering firms, as well as companies that perform engineering services, that are certified as B Corps. Some examples include US-based organisations such as Conservation Engineering, Biohabitats, and Green Dragon Engineering, as well as UK-based organisations such as E3 Consulting, Ecolife Recycling, and Greengauge Building Services. There are also firms operating in multiple countries, such as Atelier Ten, which has offices in countries such as the US, UK, Germany, and Australia.

To focus on one B Corp, Patagonia is a well-known manufacturer of outdoor apparel and equipment that is also recognised for its dedication to environmental sustainability. Patagonia's environmental efforts include campaigns to safeguard public lands and support for green businesses in its own sector. To lessen its influence on the environment, Patagonia incorporates sustainable materials in its goods, including organic cotton, recycled polyester, and help. The business also attempts to reduce waste in its operations and employs fair labour standards. Patagonia has made a commitment to decreasing its carbon emissions and has made investments in energy efficiency and renewable energy projects to do so. The business also helped form the Renewable Energy Buyers Alliance, an organisation that promotes the use of renewable energy. Through its Patagonia Action Works initiative, which awards funds and offers assistance to local environmental organisations, Patagonia promotes action within other organisations as well. Perhaps most notably, through its 'Don't buy this jacket' campaign, Patagonia urges customers to mend and reuse their apparel rather than purchasing brand-new items (Rattalino 2018).

Businesses can ensure that they are consistently taking into account both environmental and economic factors in their day-to-day operations by making sustainability a key component of their business plan. Likewise, engineering firms can make sustainability a top priority in their projects, ensuring that the projects that they work on are created with the environment in mind. For-profit firms can develop infrastructure that is more resilient to the climate crisis, more energy efficient, and constructed out of more environmentally friendly materials (e.g., timber). They can adopt green technology within their offices and in their own operations, potentially have the choice to use renewable energy sources, energy-efficient lighting, and sustainable waste management practices (Bocken et al. 2019). They can even promote environmentally friendly practices among their employees, such as advocating for carpooling, biking, and using public transport, where available. Lastly, one could argue that engineering firms have an ethical responsibility to educate their clients about the need for more sustainable designs, and the ways in which more sustainable choices can benefit the environment and save the client money.

6.3 NOT-FOR-PROFIT AND NON-PROFIT ORGANISATIONS

The terms not-for-profit and non-profit are often used interchangeably, although subtle difference exist. The key characteristic of not-for-profit and non-profit organisations is that they do not distribute profits to owners or shareholders. Not-for-profit organisations can include organisations such as cooperatives and credit units. Non-profit organisations can be considered as a subcategory of not-for-profit organisations, and they serve a specific mission or cause; examples include foundations, religious institutions, and charities.

Due to their different forms and purposes, not-for-profit organisations may take different approaches to environmental sustainability than their for-profit counterparts take. Whilst for-profit companies are generally concerned with making profits for their shareholders, not-for-profit organisations are not burdened by the need to make a profit. This opens the door to making environmental sustainability more of a priority. As discussed previously with for-profit businesses, not-for-profit organisations, especially those operating on a tight budget, can often benefit financially from adopting more sustainable behaviours. Furthermore, these sustainable behaviours and practices may align with their mission if they are focused on social and environmental issues. Not many engineering firms fall into this space, although organisations like Engineers Without Borders are not-for-profit (Helgesson 2006).

6.4 THE PUBLIC SECTOR

The public sector has a responsibility for sustainability because it is a significant part of society and the economy, and it has a significant impact on the environment. Governments have the power to enact laws and regulations that encourage the use of sustainable practices, including in transportation. They can spend money on more environmentally friendly infrastructure, including sustainable transportation like bike lanes, public transit networks, and electric vehicle charging stations, as well as renewable energy sources, such as wind, solar, and hydropower. These choices can extend to both waste management, where governments might invest in more sustainable waste management systems, and water management, where they can conserve water resources, impose regulations on water pollution, and lessen the danger of water-related natural disasters.

Governments can also set an example with their own practices, such as cutting back on their use of resources, energy, and waste. They can be a substantial influence on the public, and can encourage communities to embrace sustainability through public education and outreach campaigns. Finally, they can partner with other stakeholders, including for-profit organisations, to promote sustainability. For engineers, the public sector offers ample opportunities to shape the world more sustainably.

However, for the public sector, the balance between environmental sustainability and cost can be difficult since many sustainable programmes and activities demand sizable upfront investments. The public sector can adopt a number of actions to strike a balance between environmental sustainability and cost. Governments can give top priority to programmes that benefit the environment and are the most economical. For instance, lowering greenhouse gas emissions may be more affordable by retrofitting existing buildings with energy-efficient technologies rather than building brand-new green structures from the ground up. To support the funding of sustainable activities, governments might look into financing possibilities like public–private partnerships, green bonds, and grants from foundations, higher levels of government, and international organisations. Additionally, to make sure they are making informed decisions, governments can weigh the costs and advantages of sustainable efforts, including both immediate and long-term advantages.

REFERENCES

Anwar, Y., and N. El-Bassiouny. 2019. "Marketing and the Sustainable Development Goals (SDGs): A Review and Research Agenda." *The Future of the UN Sustainable Development Goals: Business Perspectives for Global Development in 2030*: 187–207.

Arnsperger, C., and Y. Varoufakis. 2006. "What Is Neoclassical Economics? The Three Axioms Responsible for Its Theoretical Oeuvre, Practical Irrelevance and, Thus, Discursive Power." *Panoeconomicus* 53(1): 5–18.

Beder, S. 2011. "Environmental Economics and Ecological Economics: The Contribution of Interdisciplinarity to Understanding, Influence and Effectiveness." *Environmental Conservation* 38(2): 140–150.

Bocken, N., F. Boons and B. Baldassarre. 2019. "Sustainable Business Model Experimentation by Understanding Ecologies of Business Models." *Journal of Cleaner Production* 208: 1498–1512.

Daly, H. E. 1994. *For the Common Good: Redirecting the Economy Toward Community, the Environment, and a Sustainable Future*. Boston: Beacon Press.

Helgesson, C. I. 2006. "Engineers without Borders and Their Role in Humanitarian Relief." *IEEE Engineering in Medicine and Biology Magazine* 25(3): 32–35.

Honeyman, R., and T. Jana. 2019. *The B Corp Handbook: How You Can Use Business as a Force for Good*. Oakland, CA: Berrett-Koehler Publishers.

Meadows, D. H., D. L. Meadows, J. Randers, and W. W. Behrens. 2018. "The Limits to Growth." *Green Planet Blues*, 25–29. London: Routledge.

Rattalino, F. 2018. "Circular Advantage Anyone? Sustainability-driven Innovation and Circularity at Patagonia, Inc." *Thunderbird International Business Review* 60(5): 747–755.

Redclift, M. 2005. "Sustainable Development (1987–2005): An Oxymoron Comes of Age." *Sustainable Development* 13(4): 212–227.

Stubbs, W. 2017. "Sustainable Entrepreneurship and B Corps." *Business Strategy and the Environment* 26(3): 331–344.

Xue, Y., O. Caliskan-Demirag, Y. F. Chen, and Y. Yu. 2018. "Supporting Customers to Sell Used Goods: Profitability and Environmental Implications." *International Journal of Production Economics* 206: 220–232.

History
Legacy and Long-Lasting Impacts

7

7.1 'THOSE WHO DON'T LEARN FROM HISTORY . . .'

The philosopher and author George Santayana is often credited with the saying, 'Those who don't learn from history are doomed to repeat it.' Although the exact words have changed over time, the proverb's core message has stayed constant and offers a well-known message that is still often ignored. Engineers are often forward-thinking individuals who seek to innovate for the future and thus, studying history may not seem to be so important. However, past engineering achievements and failures offer priceless lessons and insights that can guide the design and implementation of future infrastructure projects. Civil engineers can learn about the consequences of design choices, construction methods, and material selections through examining past projects, and then apply this knowledge to prevent errors and enhance project outcomes. Additionally, past engineering successes, and especially failures, can not only serve as sources of inspiration for fresh concepts but also serve as reminders of the importance of safety, ethics, and environmental sustainability in engineering practice. Civil engineers can learn valuable lessons that can guide their practice, helping to deepen their understanding of the social, economic, and environmental contexts of engineering.

DOI: 10.1201/9781003392194-7

Presented as case studies in an engineer's formal education, it may not have been apparent that what was being studied was, in fact, history. Certainly every modern civil engineer learned about the cutting-edge methods and technologies that were used to build a waterway across a tropical jungle in the early 20th century – that is, of course, the Panama Canal. Today's civil engineers will also have learned about the Burj Khalifa, and how issues such as wind loads, structural stability, and vertical transportation were addressed – and perhaps also were taught about the value of creativity and teamwork in major construction projects. Any number of other case studies of successful projects may have been presented, depending on where the engineer was educated and the breadth of exposure to projects around the globe, including the Golden Gate Bridge, the Hoover Dam, and the Big Dig in Boston (all US projects); the Thames Barrier, the Channel Tunnel, and the London Olympic Park (all UK projects); as well as the Tokyo Skytree (Japan), the Hong Kong–Zhuhai–Macau Bridge (which connects, Hong Kong, Macau, and mainland China), and the Petronas Twin Towers (Malaysia).

Historical mistakes and miscalculations also present obstacles to achieving the Sustainable Development Goals. Many of the problems we are currently facing have their roots in earlier choices, laws, and actions that had unfavourable social, economic, and environmental effects. For instance, colonialism, slavery, and other forms of exploitation and discrimination have had a lasting impact on civilisations and have fuelled inequities that still exist today. A number of SDGs are in jeopardy due to environmental degradation and the climate crisis, which have been caused, at least in part, by unsustainable practices in agriculture, industry, and energy production.

We can build a more just and sustainable future by acknowledging and attempting to correct past mistakes. This calls for an admission of the harm brought on by prior deeds and the implementation of measures to atone for it, such as through reparations, restitution, and reconciliation. It also calls for learning lessons from the past and making sure that present laws and practices adhere to the SDGs' core values of sustainability, inclusivity, and social justice.

7.2 CONFRONTING PAST MISTAKES IN ENGINEERING

Technical mistakes – that is, mistakes in the design and calculations – are often the easiest ones to highlight. Notable examples include: the collapse of the Tacoma Narrows Bridge in 1940 as a result of resonance brought on by strong winds (Arioli and Gazzola 2013); the failure of the St. Francis Dam (north of Los Angeles) in 1928 that led to a devastating flood and over 400 fatalities (Rogers 2006); the collapse of Kansas City's Hyatt Regency walkway in 1981 where 114 people died and over 200 others were injured (Hauck 1983); the failure of the levees in and around New Orleans during Hurricane Katrina in 2005 that led to massive flooding and widespread fatalities (Jonkman et al. 2009); the failure of the Oroville Dam spillway in California in 2017 that forced over 188,000 people to evacuate (Hollins et al. 2018); and the collapse of the Morandi Bridge in Genoa, Italy, in 2018 that claimed 28 lives and called into question the safety of deteriorating infrastructure (Calvi et al. 2019).

To focus on just one of those disasters, the Oroville Dam spillway disaster was a significant civil engineering failure. The catastrophe of the Oroville Dam, the tallest dam in the United States, began when torrential rains filled the reservoir behind the dam to capacity, forcing water to be released from the reservoir through the main spillway. But as a result of erosion, a sizable hole appeared in the spillway and quickly started to widen. This raised questions about the stability of the dam and forced the evacuation of downstream residents. For the first time in the dam's history, a secondary emergency spillway was opened whilst water continued to pour through the damaged spillway. Authorities were concerned that the uncontrolled flow of water over the emergency spillway could result in catastrophic flooding, but it was also discovered that this spillway was at risk of failing owing to erosion. In an effort to shore up the emergency spillway and to stop additional erosion, crews utilised helicopters to drop rocks and other materials onto it. After many days of emergency repairs, the situation at the dam was eventually stabilised, and the evacuees were able to return home. The disaster, however, brought into question the safety of aging infrastructure and the necessity of ongoing maintenance and inspection. It also emphasised the need for communication and emergency planning. After the failure, the spillway and other parts of the dam underwent extensive renovations to prevent future failures, costing in excess of a staggering $1 billion (Hollins et al. 2018, Koskinas et al. 2019, Henn et al. 2020).

Mistakes like those made with the Oroville Dam and the other examples mentioned previously are easier to acknowledge because they underscore the need for accurate and reliable data and designs. These are mistakes that should have been caught, could have been prevented, and can be guarded against repeating. These failures have led to a greater emphasis on safety and risk management, and engineers and contractors working on building projects are increasingly held to higher standards of accountability.

However, there are also mistakes that arise out of a lack of understanding of context or, worse, a lack of concern for that context. The technical aspects of developing infrastructure are typically what are emphasised by civil engineers, since they are the relevant technical experts, but this has occa-sionally resulted in overlooking or ignoring the social and cultural context in which the infrastructure is being developed. In the mid-to-late 20th cen-tury in the United States, highway construction projects frequently disrupted, and sometimes destroyed, low-income and minority neighbourhoods. The justification was that the most direct, least expensive route was the rational choice for constructing new highways or expanding existing ones, but a lack of attention was paid to the effects on the urban neighbourhood through which these highways would run. The construction of numerous highways through these settlements resulted in the eviction of local residents, the destruction of small businesses, and the disruption of social and cultural communities (Mohl 2004, Mohl 2008, Karas 2015).

Additionally, the rights and demands of indigenous populations and their connection to the environment have historically been disregarded when building water management systems like dams, canals, and pipelines. Long-standing cultural and ecological systems have been destroyed as a result of this. Large dam construction projects in particular have frequently disregarded the negative social and environmental effects of uprooting local communities. For instance, the building of the Three Gorges Dam in China resulted in the eviction of over a million people and the submerging of numerous significant historical and cultural sites (McDonald-Wilmsen 2009).

7.3 WHY THE PAST WON'T GO AWAY – AND WHY THAT'S GOOD

Past engineering mistakes continue to haunt us in the present because the public remembers mistakes that caused human loss of life, environmental harm, and monetary losses. In certain marginalised communities where these impacts have been disproportionately greater than in the larger city or region, people may be particularly sceptical and distrustful about projects that could once again cause a negative impact on them. However, future generations of engineers can learn from mistakes made by earlier generations. These makes can serve as a warning of what can go wrong and assist us in staying away from making the same errors again. The built environment has been signifi-cantly impacted throughout history by colonialism, racism, and inequity. These elements continue to affect access to opportunities and infrastructure. Civil engineers can design projects that encourage social inclusion and lessen inequities by understanding the historical context of infrastructure develop-ment. The SDGs give civil engineers a framework for making a contribution to a more just and sustainable built environment since they place a strong emphasis on eradicating poverty, advancing gender equality, and eliminating inequities.

In addition, engineering is always innovating and developing new methods of operation and materials. Engineers may create better and safer designs by learning from past mistakes and developing a better understanding of the dangers and constraints associated with new technologies and materials. Studying the past, including past mistakes, can also lend perspective – offering a more comprehensive understanding of current context by learning about how problems were overcome before, especially when those problems (e.g., greater demands for water or transport) are comparable to those we face now.

Finally, we can develop a sense of identity, community, and belonging by learning about what our predecessors have accomplished. This can strengthen our sense of purpose in engineering. For example, there is a strong legacy of water management and land reclamation in the Netherlands, allowing the nation to exist below sea level and reclaim substantial tracts of land from the sea. Afsluitdijk, a 20-mile-long dam that separates the North Sea from the freshwater Lake Ijsselmeer, and the Delta Works, a collection of dams and barriers to prevent flooding in the Netherlands, are two examples of the engineering achievements that make up this heritage (Janssen et al. 2014, Bloemen et al. 2019, Janssen et al. 2020). Additionally, the Van Brienenoordbrug and the Moerdijkbrug are two prominent examples of modern suspension bridges that were built by the Dutch (Buitelaar et al. 2004, Bliemer et al. 2009). The Port of Rotterdam, one of the greatest ports in the world and a significant hub of international trade for centuries, is located in the Netherlands. The legacy of Dutch engineering is still evident today, with the Netherlands' first solar road and the Amsterdam electric bus project serving as examples of the nation's dedication to sustainability and innovation (Bakker and Konings 2018, Shekhar et al. 2018). The Dutch are renowned for their creative approaches to energy and transportation – and an understanding of this history can help to inspire future engineers.

REFERENCES

Arioli, G., and F. Gazzola. 2013. "Old and New Explanations of the Tacoma Narrows Bridge Collapse." Atti XXI Congresso AIMETA, Torino.

Bakker, S., and R. Konings. 2018. "The Transition to Zero-emission Buses in Public Transport – The Need for Institutional Innovation." *Transportation Research Part D: Transport and Environment* 64: 204–215.

Bliemer, M., M. Dicke-Ogenia, and D. Ettema. 2009. "Rewarding for Avoiding the Peak Period: A Synthesis of Three Studies in the Netherlands." European Transport Conference 2009, Citeseer.

Bloemen, P., M. Van Der Steen, and Z. Van Der Wal. 2019. "Designing a Century Ahead: Climate Change Adaptation in the Dutch Delta." *Policy and Society* 38(1): 58–76.

Buitelaar, P., R. Braam, and N. Kaptijn. 2004. "Reinforced High Performance Concrete Overlay System for Rehabilitation and Strengthening of Orthotropic Steel Bridge Decks." Orthotropic Bridge Conference, Sacramento, USA.

Calvi, G. M., M. Moratti, G. J. O'Reilly, N. Scattarreggia, R. Monteiro, D. Malomo, P. M. Calvi, and R. Pinho. 2019. "Once Upon a Time in Italy: The Tale of the Morandi Bridge." *Structural Engineering International* 29(2): 198–217.

Hauck, G. F. 1983. "Hyatt-Regency Walkway Collapse: Design Alternates." *Journal of Structural Engineering* 109(5): 1226–1234.

Henn, B., K. N. Musselman, L. Lestak, F. M. Ralph, and N. P. Molotch. 2020. "Extreme Runoff Generation from Atmospheric River Driven Snowmelt during the 2017 Oroville Dam Spillways Incident." *Geophysical Research Letters* 47(14): e2020GL088189.

Hollins, L. X., D. A. Eisenberg, and T. P. Seager. 2018. "Risk and Resilience at the Oroville Dam." *Infrastructures* 3(4): 49.

Janssen, S., H. Vreugdenhil, L. Hermans, and J. Slinger. 2020. "On the Nature Based Flood Defence Dilemma And its Resolution: A Game Theory based Analysis." *Science of the Total Environment* 705: 135359.

Janssen, S. K., A. P. Mol, J. P. Van Tatenhove, and H. S. Otter. 2014. "The Role of Knowledge in Greening Flood Protection. Lessons from the Dutch Case Study Future Afsluitdijk." *Ocean & Coastal Management* 95: 219–232.

Jonkman, S. N., B. Maaskant, E. Boyd, and M. L. Levitan. 2009. "Loss of Life Caused by the Flooding of New Orleans after Hurricane Katrina: Analysis of the Relationship between Flood Characteristics and Mortality." *Risk Analysis: An International Journal* 29(5): 676–698.

Karas, D. 2015. "Highway to Inequity: The Disparate Impact of the Interstate Highway System on Poor and Minority Communities in American Cities." *New Visions for Public Affairs* 7: 9–21.

Koskinas, A., A. Tegos, P. Tsira, P. Dimitriadis, T. Iliopoulou, N. Papanicolaou, D. Koutsoyiannis, and T. Williamson. 2019. "Insights into the Oroville dam 2017 Spillway Incident." *Geosciences* 9(1): 37.

McDonald-Wilmsen, B. 2009. "Development-induced Displacement and Resettlement: Negotiating Fieldwork Complexities at the Three Gorges Dam, China." *The Asia Pacific Journal of Anthropology* 10(4): 283–300.

Mohl, R. A. 2004. "Stop the Road: Freeway Revolts in American Cities." *Journal of Urban History* 30(5): 674–706.

Mohl, R. A. 2008. "The Interstates and the Cities: The US Department of Transportation and the Freeway Revolt, 1966–1973." *Journal of Policy History* 20(2): 193–226.

Rogers, J. D. 2006. *Lessons Learned from the St. Francis Dam Failure*. Reston, Virginia, USA: Geostrata.

Shekhar, A., V. K. Kumaravel, S. Klerks, S. de Wit, P. Venugopal, N. Narayan, P. Bauer, O. Isabella, and M. Zeman. 2018. "Harvesting Roadway Solar Energy – Performance of the Installed Infrastructure Integrated PV Bike Path." *IEEE Journal of Photovoltaics* 8(4): 1066–1073.

Public versus Private Space

8

Whose Bridge Is It?

8.1 WHY PUBLIC AND PRIVATE SPACE MATTER FOR THE UNITED NATIONS' SUSTAINABLE DEVELOPMENT GOALS

Both public and private space matter in achieving the UN's Sdgs, but the roles that public and private space play in achieving those goals often differs. SDG 11 (Sustainable Cities and Communities) sets a goal of providing secure, welcoming, and accessible green space as well as public space more generally. The promotion of sustainable cities and communities depends heavily on access to public places like parks, walkways, and bike lanes, which can support social inclusion, support sustainable and active transport modes, and promote a sense of local ownership and responsibility.

SDG 3 (Good Health and Well-Being) seeks to promote health, and parks, recreation centres, and sports venues are examples of places that encourage physical activity and good health. The achievement of SDG 3 – healthy lifestyles and well-being for all – relies heavily on public spaces such as parks, which tend to be public, although recreation centres and sports venues may be either public or private.

Additionally, public and private spaces are frequently gendered, including how women tend to experience more barriers to accessing public spaces (Franck and Paxson 1989, Garfias Royo et al. 2020). SDG 5 (Gender Equality) seeks to empower women to achieve gender equality, in part through increasing access to public amenities and services. Private spaces (such as homes, companies, and schools) also play a significant role; for example, having a

 DOI: 10.1201/9781003392194-8

place to live that is safe and secure can enable women and girls to fully engage in society and to accomplish their goals.

8.2 WHY PUBLIC SPACE MATTERS

Public spaces are any areas that are open and access to members of the general public. They can be found inside and outside and can include spaces such as parks, plazas, sidewalks/footpaths, streets, public buildings (such as libraries, museums, and community centres), and other places of a similar nature. They are often owned, run, or maintained by government (whether town/city, national, or another level) or non-profit institutions. Public spaces are crucial in fostering inclusive and inviting communities that foster a sense of connection and belonging (Carmona 2019).

Cities are often known by their iconic landmarks that, in some cases, are public. The Eiffel Tower in Paris, the Golden Gate Bridge in San Francisco, and the Plaza de Mayo in Buenos Aires are all public (although the Eiffel Tower does require a paid ticket to visit). When designing the built environment, and especially elements of the public realm like bridges and civic buildings, one is designing the very look and feel of a city or town. Urban areas' aesthetic appeal can also be improved through a thoughtful redesign of public spaces. For example, sculptures and installations of public art can improve the aesthetic appeal of plazas and parks and increase people's enjoyment of them. Well-designed public spaces can support the economy of a community by bringing in tourists, supporting small businesses, and raising property values.

Public space serves a variety of social, cultural, environmental, and economic purposes that are necessary for achieving resilient and sustainable communities. Public spaces offer room for social interaction and cultural exchange where people from various backgrounds can meet, connect, and foster a sense of community. People can meet with friends and make new acquaintances in public areas. Exhibitions, fairs, and other cultural events can all take place in public areas. Through these kinds of cultural events, people can learn about and develop an appreciation for the various cultures that make up their community.

Public areas can also serve a civic purpose. They are frequently used as locations for neighbourhood gatherings, open forums, and political rallies and demonstrations. History is replete with examples of large-scale protests and strikes in public squares and parks, from Tiananmen Square (Beijing, 1989), to the original Occupy Wall Street protests in Zuccotti Park (New York City, 2011) to the Tahrir Square Protests (Cairo, 2011). Even though plazas and parks have sometimes been the site of conflict and even bloodshed, they serve an important democratic and public purpose in providing a space for demonstration and protest. They can encourage civic involvement and provide people a way for taking part in their communities by offering a forum for public discourse. In times of emergency, such as natural catastrophes, public spaces can also serve as gathering places and distribution points for emergency supplies.

Public areas can promote physical activity and improve mental well-being (Koohsari et al. 2015). Opportunities for physical and recreational activity are abundant in places like parks, playgrounds, and outdoor sporting venues. By encouraging a more active lifestyle for community members and providing space for people to unwind, meditate, and rest, these areas can improve the physical and mental health of the community. Studies have shown that spending time in green areas like parks and gardens can help lower stress levels and increase happiness (Grahn and Stigsdotter 2010). Additionally, public infrastructure such as sidewalks/footpaths, bicycle lanes and paths, multiuse paths through parks can encourage physical activity, both as the primary trip mode as well as walking/biking in conjunction with taking public transit (Terzano and Morckel 2011). Public transport hubs, such as Grand Central Terminal in New York City and Shinjuku Station in Tokyo, can be important junctions for transport both within cities and between cities.

Last but not least, public areas promote environmental sustainability. Public areas, especially parks, can offer much-needed green space in areas where access to nature could be limited. Due to the dense concentration of buildings and paved surfaces, cities often experience a heat island effect whereby urban areas are warmer than neighbouring rural areas. Green spaces can help to offset this effect. Parks can also provide habitat for a variety of plants, animals, and insect, aiding in the preservation of biodiversity. These habitats, which might not be present elsewhere in urban settings, can maintain populations that might otherwise be endangered. Parks and tree-lined streets also remove carbon dioxide and other greenhouse gases from the air and release oxygen through photosynthesis, improving air quality. Trees can provide shade, which can cool an area and lessen the need for air conditioning in nearby buildings, as well as reduce stormwater flow by absorbing water and serving as a control for erosion.

8.3 WHY PRIVATE SPACE MATTERS

Within a city, private space refers to spaces that are privately owned and solely utilised by people or groups for their own needs or those of their businesses. A city's residences, gated communities, commercial structures, and most workplaces are examples of private areas. Private spaces are governed by the policies established by their owners or landlords, as opposed to public spaces, which are owned and managed by the government or other public institutions. This can involve setting limits on who is allowed to use the area, how it is used, and what kinds of activities are allowed there (Landman 2006).

Privately owned areas provide individuals and organisations with independence and control over their environment. Property rights are often regarded as essential to individual freedom and economic growth in many nations, going as far back as at least the 17th century and John Locke's *Two Treatises of Government*. These property rights are acknowledged and upheld by the government, enabling people and businesses to own land and structures and to utilise them as they see fit.

The built environment can benefit from privately owned spaces. Buildings and other structures can be purchased using funds that private individuals and businesses can use for job creation and economic expansion. Through private ownership, there is also arguably a greater variety of design, as private owners can design and alter their structures to accommodate their unique requirements and preferences, creating a more varied and interesting urban environment. For better or worse, profit often drives private owners, encouraging them to use their assets (including land and buildings) as productively and efficiently as they can.

On an individual level, people value private space for the personal space and privacy afforded to them. Finding peaceful and isolated spaces in densely populated urban areas can be challenging, and people can escape the commotion of city life and unwind in their own private areas, such as their houses, flats, and gardens. Additionally, private spaces can afford physical security – shelter from the weather and a secure place to store personal belongings – as well as a financial security. Within their own spaces, people are also free to express their individual preferences and tastes, reflecting their own style and personality in their home furnishings and landscaping.

8.4 THE TENSION BETWEEN PUBLIC AND PRIVATE SPACE

Tension or conflict can develop when people misuse or abuse public areas (e.g., through littering or loud music), or when people use private spaces in ways that have detrimental effects on the public realm (Németh and Schmidt 2011). For example, a factory or power plant may release pollutants that impact the surrounding area's air and/or water quality. A developer might construct a tall building that obstructs the view of neighbouring open space or casts shadows onto this space. In these situations, how private space is built and used may have broader effects on the neighbourhood (Madanipour 1999).

Private spaces are created for the use, enjoyment, and benefit of the owner (whether an individual or a company), whereas public spaces are meant to be shared resources that serve the needs of the community. Tension between the public and private sections can arise when people or organisations use public spaces in ways that harm or take away from the common good, or when private areas are perceived to be harming the neighbourhood. It is critical for people, organisations, and governments to accept accountability for their actions and work towards solutions that advance both the public and private realms in order to reduce these tensions.

Certain parts of the built environment belong quite obviously to either the public or private realm, such as a city hall, which is clearly public, or a personally financed home, which is clearly private. However, sometimes it is unclear whether a structure or a plot of land is public or private, or even something that could be termed privately owned public space (Kayden 2000). For example, bridges are usually part of the public realm that are built and maintained by governments, but a number of bridges worldwide are run and controlled by private organisations or people (Fisher 2014). Bridges may become privately owned for a number of different reasons. A private corporation might construct a bridge as a component of a broader infrastructure project, such as with toll roads. Alternatively, a person or group may decide to invest by buying an existing bridge.

Bridges that are privately owned may have benefits or drawbacks. On the one hand, since the owner has a financial stake in the bridge's performance, private ownership can encourage effective operation and maintenance of the structure. Since the owner may have more latitude in decision-making and funding, private ownership might also facilitate more innovative design and construction techniques. On the other hand, since the owner may impose tolls or other fees that make it challenging for some people to use the bridge, private ownership can also give rise to concerns about accessibility and affordability. Since the government may have little control over the conduct of private owners, private ownership can also present difficulties in terms of oversight and regulation. Thus, while private ownership of bridges is a possibility, it is crucial to weigh the potential advantages and disadvantages in each unique situation to make sure that the interests of the public and the private owner are balanced.

Parks are another case where it can be ambiguous whether they are public or private, even though they are generally public. Bryant Park in New York City is managed by the Bryant Park Corporation, which is a private non-profit organisation, but the park remains public (Schmidt and Németh 2010). The park was formerly a piece of land owned by the New York Public Library, but in the 1970s it fell into disrepair and earned a reputation as an unsafe and unwelcoming place. The Bryant Park Corporation, which was established in 1980 to manage the park, and the City of New York collaborated to revitalise the space (Green 2022). According to the partnership's agreements, the city owns the land and the park, while the Bryant Park Corporation is in charge of its regular management, upkeep, and programming. The city, businesses that sponsor the park, as well as donations and other contributions from the public and private sectors all contribute to the organisation's funding (Madden 2010).

When New York Fashion Week was hosted in Bryant Park, annually between 1993 and 2009, there was some criticism around it. Fashion Week, now held elsewhere, is a private event but was being held in a public park. Critics voiced a concern that the park was effectively being privatised because so much of the park was blocked off to the general public and utilised only for fashion industry activities during the festival. This led to concerns about whether the general public was being excluded from a location that is meant to be accessible to everybody.

There were also concerns about how Fashion Week was affecting the park, which included concerns that the event was seriously damaging the grass and other landscaping elements as well as the park's infrastructure, such as its paths and seating areas. Concerns were raised about the event's crowds and noise, which some local stakeholders believed interrupted the neighbourhood's regular operations. Finally, the public, according to critics, was being made to pay for upkeep and repairs of a park that was being damaged during a private event.

Supporters of the event countered that Fashion Week helped to reinforce New York's image as a major global centre of fashion and that it was a significant cultural and economic event for the city. They also emphasised that the City of New York was in charge of regulating and monitoring the event and that precautions were being taken to mitigate any negative effects on the park and its surroundings. Overall, there was discussion and controversy about the use of Bryant Park for Fashion Week, with arguments made on both sides regarding whether it was appropriate to use a public area for a private event. Fashion Week moved on to be held in various other venues, including, since 2019, a creative events space called Spring Studios, which is privately owned (Green 2022).

REFERENCES

Carmona, M. 2019. "Principles for Public Space Design, Planning to do Better." *Urban Design International* 24: 47–59.
Fisher, T. 2014. "Public Value and the Integrative Mind: How Multiple Sectors can Collaborate in City Building." *Public Administration Review* 74(4): 457–464.

Franck, K. A., and L. Paxson. 1989. "Women and Urban Public Space: Research, Design, and Policy Issues." *Public Places and Spaces*: 121–146.

Garfias Royo, M., P. Parikh, and J. Belur. 2020. "Using Heat Maps to Identify Areas Prone to Violence against Women in the Public Sphere." *Crime Science* 9(1): 1–15.

Grahn, P., and U. K. Stigsdotter. 2010. "The Relation between Perceived Sensory Dimensions of Urban Green Space and Stress Restoration." *Landscape and Urban Planning* 94(3–4): 264–275.

Green, T. 2022. *New York Fashion Week Review: A Look at Trends from 1950 to 2020.* San Marcos, Texas, USA: Texas State University.

Kayden, J. S. 2000. *Privately Owned Public Space: The New York City Experience.* Hoboken: John Wiley & Sons.

Koohsari, M. J., S. Mavoa, K. Villanueva, T. Sugiyama, H. Badland, A. T. Kaczynski, N. Owen, and B. Giles-Corti. 2015. "Public Open Space, Physical Activity, Urban Design and Public Health: Concepts, Methods and Research Agenda." *Health & Place* 33: 75–82.

Landman, K. 2006. "Private Space-'private Citizen': What Kind of Cities Are We Creating?: Seminar Proceedings: Gated Communities." *SA Publiekreg= SA Public Law* 21(1): 51–71.

Madanipour, A. 1999. "Why are the Design and Development of Public Spaces Significant for Cities?" *Environment and Planning B: Planning and Design* 26(6): 879–891.

Madden, D. J. 2010. "Revisiting the End of Public Space: Assembling the Public in an Urban Park." *City & Community* 9(2): 187–207.

Németh, J., and S. Schmidt. 2011. "The Privatization of Public Space: Modeling and Measuring Publicness." *Environment and Planning B: Planning and Design* 38(1): 5–23.

Schmidt, S., and J. Németh. 2010. "Space, Place and the City: Emerging Research on Public Space Design and Planning." *Journal of Urban Design* 15(4): 453–457.

Terzano, K., and V. C. Morckel. 2011. "Walk or Bike to a Healthier Life: Commuting Behavior and Recreational Physical Activity." *Environment and Behavior* 43(4): 488–500.

The People Working with and on Behalf of the Public

9

9.1 THE ETHICAL RESPONSIBILITY TO SERVE THE PEOPLE

Serving the public is a significant ethical responsibility for civil engineers. Civil engineering projects directly affect the general public's safety, health, and well-being. Because of this, civil engineers must take care to ensure that they perform their work in a way that puts public welfare and safety first.

The UK's professional association for civil engineers, the Institution of Civil Engineers (ICE), has created a Code of Ethics that outlines the ethical responsibilities of civil engineers. The code places a strong emphasis on the value of accountability, honesty, and integrity in all facets of engineering work. Additionally, it emphasises the duty of civil engineers to safeguard the environment, advance sustainability, and guarantee that their work serves society as a whole.

One of the main ethical obligations of civil engineers is to make sure that their work complies with all relevant rules and regulations. This involves making sure that their designs and constructions will not endanger the health or safety of the general public. Effective communication with the public and other stakeholders is an additional ethical responsibility. This entails giving the public clear and accurate information on engineering projects, paying attention to their comments and concerns, and resolving any issues or complaints that may come up. Additionally, it is the ethical responsibility of civil engineers to ensure that the infrastructure that they design and construct satisfies the needs of the community it serves. This involves consideration for accessibility as well as creating infrastructure that is both useful and aesthetically pleasing.

This ethical responsibility relates to at least three of the UN's Sustainable Development Goals. By planning and constructing safe, reliable, and sustainable infrastructure, civil engineers support SDG 9: Industry, Innovation, and Infrastructure. In addition to constructing infrastructure that can resist natural disasters and other hazards, this also means using technologies and materials that can lessen the environmental impact of construction. By designing infrastructure that is user-friendly, accessible, and satisfies community requirements, civil engineers contribute to achieving SDG 11: Sustainable Cities and Communities. This includes creating green spaces that support biodiversity and enhance quality of life, as well as building transport systems that alleviate traffic congestion and reduce pollution.

Finally, by creating infrastructure that is resilient to the effects of the climate crisis and that aids in the transition to a low-carbon economy, civil engineers support SDG 13: Climate Action. This involves creating infrastructure for renewable energy sources and creating buildings with energy-efficient features that lower greenhouse gas emissions.

9.2 COMMUNITY STAKEHOLDERS

Community stakeholders are people or organisations with an interest or investment in a certain community. They may consist of: residents (people who live in the community and are affected by its development and changes within it), business owners (those who own or run businesses in a community), non-profit organisations (groups working to improve the community through social services, education, and advocacy), local government (elected officials and public servants who are responsible for making decisions that affect the community), community groups (groups of residents who organise informal groups to address community issues), and educational institutions (at the primary, secondary, and tertiary level, and including both public and private institutions).

Each type of community stakeholder has a valuable contribution to make because they all have different perspectives, resources, and areas of expertise. Depending on the situation and the challenges involved, a particular stakeholder group may or may not significant. For example, if a community is experiencing a financial crisis, business owners may be particularly significant stakeholders because their performance is closely related to the community's economic health. On the other hand, the local government and healthcare professionals may be more significant stakeholders in a town experiencing a public health crisis.

Residents are often seen as the most significant community stakeholders because they live, and often work, in the community and directly impact its success or failure. To make sure that their opinions are heard and that their needs are met, it is crucial to involve and empower individuals in local decision-making processes. However, it is important to value the viewpoints and contributions of each stakeholder group since effective community relations and, thus, successful projects, ultimately depend on the participation and cooperation of all stakeholders.

9.3 DIVERSE AND DIVERGENT OPINIONS

Due to their varied backgrounds, experiences, and potential for differences in priorities, community stakeholders may have a variety of differing and contrasting perspectives. For example, in a community development project, certain stakeholders may place a higher priority on job creation and economic growth than on environmental sustainability. Furthermore, stakeholders' priorities and opinions may differ depending on their beliefs, values, and cultural perspectives.

Limited resources can also significantly contribution to friction and conflict amongst community stakeholders. Stakeholders may compete for access to scarce resources, such as money, land, or other assets, in order to meet their own demands or realise their desired goals. In areas where there is a shortage of land for housing, various parties may have opposing views on how the land should be used, resulting in conflicts between developers, local residents, and environmentalists. Conflicts, disagreements, and occasionally more serious forms of tension like violence can result from this competition (Johnson and Klassen 2022).

Water resources are another example of how a lack of resources can lead to tension and disputes amongst community stakeholders. Water is a limited resource in many areas, and stakeholders may have conflicting demands for its use. Agriculture can require significant amounts of water for irrigation, whereas urban areas also demand water for residential and industrial uses. This can lead to a conflict between farmers and urban stakeholders. Urban inhabitants may believe that farmers are using too much water for irrigation whilst farmers may believe that water is being unfairly diverted to urban uses, resulting in a lack of water for rural areas (Hargrove and Heyman 2020).

Stakeholders may also have differing degrees of influence, power, and authority within the community. Some may hold formal positions of authority, such as those of elected politicians or business owners. Others may be members of the community who lack formal authority but have extensive first-hand knowledge and experience within the local environment. Divergent attitudes amongst community stakeholders can be greatly influenced by these power dynamics. Power is the capacity to direct resources, influence opinions, and shape decision-making. The outcome of a community initiative or decision may be more heavily influenced by those with more power, whilst others with less power may feel ignored or marginalised.

For example, a business owner might be more powerful and influential in an economic development initiative than a low-income resident or a supporter of environmental causes. The owner of the business might put economic development and profit first, whereas a low-income person may have different priorities, such as affordable housing and social services, and an environmental advocacy group is likely to put environmental sustainability first.

9.4 RECONCILING CONFLICTING GOALS AND DESIRES

Tension, conflicts, and differences of opinion amongst stakeholders may result from community stakeholders holding differing points of view, differing levels of power, and differing goals and desires. Those in positions of power may control decision-making processes without the use of appropriate community engagement and communication tactics, whilst those stakeholders with less power may feel excluded or unheard. This may erode stakeholder cooperation and trust, limiting the project's potential benefits for the whole community.

To accomplish shared objectives and achieve results, it is crucial to accept and acknowledge divergent viewpoints as well as find solutions to conflicts. It is also critical to acknowledge power disparities and guarantee that all parties have a say in decision-making. This can involve establishing inclusive and open channels for participation and communication, supplying information and resources to assist stakeholders in comprehending complicated or technical information, and establishing chances for discussion and collaboration to foster confidence and consensus. Good community and community involvement tactics can support this process and guarantee that all interested parties have a role in decision-making.

Reconciling divergent perspectives may also involve enacting laws and regulations that encourage the fair allocation and effective use of resources; for example, with regard to water, there may need to be water pricing schemes, campaigns to conserve water, or low-cost distribution of water-saving devices.

The Dakota Access Pipeline project in the United States is a well-known example of community stakeholder conflict over infrastructure. The Missouri River would be crossed by the pipeline as it travelled from North Dakota to Illinois, passing by the Standing Rock Sioux Reservation in North Dakota. The Standing Rock Sioux Tribe and their allies expressed concern over the pipeline's impacts on their holy places, cultural heritage, and water supply. They asserted that they had not been appropriated consulted or able to participate in the decision-making process, and that the pipeline would threaten the environment and public health (Whyte 2017).

Activists and law enforcement were embroiled in an eight-month-long protest and series altercations over the construction of the pipeline, which culminated in a standoff at the Standing Rock protest camp in 2016–2017 (Goeckner et al. 2020). The demonstrations increased awareness of the problem on a national and international level, bringing attention to the conflict between environmental preservation, Native populations' rights, and economic growth, as well as the need for inclusive and participatory decision-making processes. The Dakota Access Pipeline example highlights how crucial it is to involve all relevant parties in infrastructure development projects, especially those who have a history of being marginalised and those who would be most impacted.

Stakeholder involvement, when executed correctly, may help with problem-solving, trust-building, and identifying and addressing issues that are of concern to all parties (Dvarioniene et al. 2015). However, it is also important to remember that there are situations where conflicting interests may not be able to be reconciled – and the Dakota Access Pipeline was likely one of those situations. This irreconciliation may arise because stakeholders cannot compromise on their fundamentally differing values, priorities, or goals. In some circumstances, the interests of the various stakeholders may be so dissimilar that it is impossible to come up with a solution that will appease everyone. Stakeholders may have irreconcilable viewpoints, for example, when the protection of natural resources and economic development are at odds. Additionally, it may be challenging or impossible to reconcile conflicting interests in circumstances where there are power imbalances, such as in cases of systemic discrimination or marginalisation. Finding a solution that is equitable and inclusive may be difficult when the interests of dominant groups and those of marginalised groups conflict (Fuller 2006).

Although collaboration and consensus amongst stakeholders should be the goal, it is also critical to acknowledge that there may be circumstances in which competing interests cannot be balanced. In those situations, facilitators need to ensure that they hold respectful discussions, look for areas of agreement when possible, and base judgements on the best facts and data at hand.

Facilitators and project leaders are also cautioned against appearing to favour the dominant group over marginalised groups. Favouring the dominant group would go against the values of fairness, equity, and human rights, whilst also perpetuating systematic discrimination, inequality, and injustice. Additionally, there is evidence that the most marginalised populations of the world are suffering the most from the effects of the climate crisis (Sibiya et al. 2022). All interested parties, especially disadvantaged and underrepresented groups, must participate for decision-making to be inclusive and equitable. Power disparities, prejudice, and historical injustices that have resulted in the marginalisation of particular communities must be acknowledged and addressed. By doing so, we are able to build more inclusive, just societies that uphold and defend human rights and the dignity of every person. Additionally, even the appearance of favouring the dominant group over the marginalised minority can erode stakeholder trust and collaboration, and lead to social unrest, political instability, and economic inefficiencies. On the other hand, marginalised groups can contribute unique viewpoints, information, and talents that can result in more creative and efficient solutions to complicated problems when they are included and given authority.

Finally, the UN's SDGs are founded on the idea of 'leaving no one behind,' which highlights the importance of addressing the needs and aspirations of the most vulnerable and disadvantaged populations. SDG 16 seeks to advance inclusive and peaceful societies, and SDG 17 highlights the need for partnerships and collaboration. The SDGs aim to create a more equitable and sustainable society that is inclusive of all people and communities by giving priority to the needs of disadvantaged groups and aiming to eradicate the causes of prejudice and inequality.

REFERENCES

Dvarioniene, J., I. Gurauskiene, G. Gecevicius, D. R. Trummer, C. Selada, I. Marques, and C. Cosmi. 2015. "Stakeholders Involvement for Energy Conscious Communities: The Energy Labs Experience in 10 European Communities." *Renewable Energy* 75: 512–518.

Fuller, B. 2006. *Trading Zones: Cooperating for Water Resource and Ecosystem Management When Stakeholders Have Apparently Irreconcilable Differences.* Cambridge, MA: Massachusetts Institute of Technology.

Goeckner, R., S. M. Daley, J. Gunville, and C. M. Daley. 2020. "Cheyenne River Sioux Traditions and Resistance to the Dakota Access Pipeline." *Religion and Society* 11(1): 75–91.

Hargrove, W. L., and J. M. Heyman. 2020. "A Comprehensive Process for Stakeholder Identification and Engagement in Addressing Wicked Water Resources Problems." *Land* 9(4): 119.

Johnson, P. F., and R. D. Klassen. 2022. "New Directions for Research in Green Public Procurement: The Challenge of Inter-stakeholder Tensions." *Cleaner Logistics and Supply Chain* 3: 100017.

Sibiya, N., M. Sithole, L. Mudau, and M. D. Simatele. 2022. "Empowering the Voiceless: Securing the Participation of Marginalised Groups in Climate Change Governance in South Africa." *Sustainability* 14(12): 7111.

Whyte, K. 2017. "The Dakota Access Pipeline, Environmental Injustice, and US Colonialism." *Red Ink: An International Journal of Indigenous Literature, Arts, & Humanities* 19(1).

Other People Working with Architects, Planners, Construction Professionals, and More

10

10.1 THE LANDSCAPE OF ENGINEERING-RELATED PROFESSIONALS

Throughout the phases of project planning, design, building, and maintenance, civil engineers work with a range of other professionals (Chan et al. 2002, Christodoulou 2004). Architects design buildings and other structures that are practical, secure, and aesthetically pleasing. Town/urban planners collaborate with civil engineers in planning and constructing urban infrastructure and public spaces like parks, building, and transit systems. For a project's funding, approvals, and compliance with rules and codes, civil engineers may collaborate with government officials are various levels.

Surveyors measure and chart the land, and they work with civil engineers to choose the best site for a project. Material suppliers are responsible for supplying steel, concrete, asphalt, or other materials for a project. Contractors are in charge of carrying out a project's construction, and they, along with construction inspectors, make sure that the building requirements are upheld.

DOI: 10.1201/9781003392194-10

Project managers are in charge of overseeing the entire project to ensure it finishes on schedule and within budget.

Civil engineers also work with other kinds of engineers. For example, geotechnical engineers study the qualities of rock and soil at a site, whereas environmental engineers ensure that projects abide by environmental laws and are planned to have the least possible negative effects on the environment.

10.2 THE ARCHITECT'S PERSPECTIVE

Civil engineers and architects have distinct but complementary roles in the design and construction of buildings and other structures. An architect's responsibility lies in the aesthetic and functional design of a building or structure. They consider the building's intended use, the location and surroundings, and the desired aesthetic of the structure. They provide blueprints and drawings that depict the structure's design, look, and construction materials. They often collaborate with the client to ensure the design meets the client's needs (Todoroff et al. 2021).

A civil engineer is responsible for the project's technical and structural components. They are in charge of making sure a building is safe, structurally sound, and able to withstand the forces it will encounter over its lifetime. They consider the soil composition, the building's weight, and the loads it will have to support. They plan and direct the construction of the building's foundation and structural supports. A civil engineer is focused on the technical and structural elements that make a building safe and stable, whereas an architect is focused on the layout and aesthetics of the structure.

Conflicts between architects and civil engineers can occur because of their different priorities, objectives, and viewpoints. For example, conflicting design priorities may lead to friction. Architects may emphasise a building's visual appeal and usability over the civil engineer's concern for structural soundness. Conflicts over the building's design, such as the size of windows, the location of structural supports, to the choice of materials, may result from this.

Different levels of technical expertise can become a source of conflict as well, and civil engineers would do well to remember that architects have their own expertise that goes beyond merely making a building look beautiful. There may be misconceptions or miscommunications regarding the design requirements because of the differences in technical expertise, however. Similar issues may arise when civil engineers fail to fully comprehend an architect's design intent, which can result in disagreements about the viability or usefulness of particular design elements. Additionally, conflict may also result from project budgets, schedules, and building techniques. Civil engineers may prioritise certain building techniques or materials, even if it means using more expensive materials or taking longer to complete the project, whereas architects may prioritise finishing the project on schedule and within budget.

However, cooperation and respect between architects and civil engineers is essential for the completion of successful projects. Together, they can design and create buildings and structures that are both aesthetically pleasing and structurally sound.

10.3 THE PLANNER'S PERSPECTIVE

From the perspective of the town/urban planner, civil engineers lend essential technical expertise to the planning process, ensuring that the infrastructure and buildings are safe, functional, and well integrated into the surrounding environment. Planners create a conceptual plan for a development that shows how roadways, buildings, open areas, and other uses will be arranged. In order to guarantee that the design complies with engineering standards and regulations, civil engineers advise on the plan's viability and offer revisions. Civil engineers also carry out a site analysis to assess a project's feasibility based on the geotechnical, hydrological, and environmental aspects of the site, and planners receive these findings and use them to modify the project's design (Bauer 2009).

Urban planners value civil engineers' contributions to addressing problems with sewage treatment, water supply, and traffic flow, amongst other infrastructure issues. They collaborate closely with civil engineers to make sure that planned developments will satisfy community demands whilst also adhering to engineering standards and regulations, and that the infrastructure is incorporated into the development's overall design. Town planners are concerned with balancing civil engineering solutions with the social, economic, and environmental factors in a community. Planners may prioritise community demands like walkability, accessibility, and connection, as well as aesthetic issues like urban design, character, and creating a sense of place.

Through their close collaborations, planners and civil engineers can develop effective and sustainable communities that meet the needs of the people who live, work, and recreate there. Town planners view civil engineers as valued partners who share the same objective of building liveable, sustainable,

and resilient communities, and who offer technical expertise that is essential
to the planning process (Ozcep and Ozcep 2011).

10.4 THE CONSTRUCTION
PROFESSIONAL'S PERSPECTIVE

Construction professionals can include a wide range of roles from surveyors
to material suppliers to contractors, project managers, construction inspectors, and more (Hussin and Omran 2009). Successful projects rely on good
communication and collaboration with a variety of construction professionals.

Although civil engineers are likely to have performed some surveying
and field studies as part of their training, surveyors are experts at measuring and mapping the land, borders, and natural characteristics of a property.
They use surveying tools like theodolites and GPS to establish exact mea-
surements and boundary lines, which a civil engineer then uses to create an
effective design. Before a building project starts, surveyors are frequently
engaged to make sure the land is appropriate for the intended purpose and to
map out the exact property boundaries. However, if the surveyor's findings
are incorrect or insufficient, this may cause disputes, conflicts, or mistakes
later on. For example, if the survey incorrectly locates a property line, it
may result in disagreements with neighbouring property owners or have a
negative impact on the project's timeline. Similarly, if the surveyor fails to
detect critical site elements such as underground soil or water conditions, it
may lead to design faults that compromise the structure's safety or structural
integrity. In order to ensure that project is designed and built precisely and
efficiently, it is crucial for the surveyor and civil engineer go work together
and communicate clearly.

Civil engineers and contractors work together during the building process. Contractors are responsible for construction site management and coor-
dination, in accordance with the plans given by the client or civil engineer.
Contractors are responsible for duties such as acquiring the supplies needed
for the project; managing the on-site construction crews; arranging and organ-
ising the building work with subcontractors; and ensuring that the construction is finished on schedule, within budget, and in compliance with safety
standards. Civil engineers may periodically visit the building site to check on
the progress of the construction and to make sure the plans are being followed
appropriately. As problems arise or adjustments are needed throughout the
construction process, the contractor may also seek out the civil engineer. Civil
engineers and contractors have complementary roles, with the civil engineer

providing the technical expertise and design knowledge, whilst the contractor oversees the project and organises the work of the numerous trades involved. A project manager works closely with the contractor and civil engineer to ensure the project is finished on schedule, on budget, and to the necessary quality standards. The project manager is in charge of supervising the entire project from beginning to end. The project manager creates the project plan, which includes the timetable; collaborates with the civil engineer to ensure the plans and design adhere to the project's specifications; hires the contractor and sometimes the subcontractors as well; administers contracts; tracks costs, controlling cash flow, and making sure the project stays within its budgets; and monitors progress. The project manager plays a critical role in a construction project's success. They serve as the point of contact for the client, the civil engineer, and the contractor, making sure that everyone is collaborating successfully to meet the project's objectives.

The role of the construction inspector is to make sure the building site's construction work complies with all applicable standards, codes, and regulations. They execute on-site inspections of the construction work independently of the civil engineer, contractor, and project manager. The construction inspector is also responsible for finding any problems or flaws in the construction process; submitting reports to the project manager, civil engineering, and contractor after recording their findings; and examining and approving the construction work at several project stages, including the foundation, the framing, and the final inspection. The construction inspector is relied upon to offer objective and impartial assessment on the construction work.

10.5 THE ECOLOGIST'S PERSPECTIVE

When working together to create infrastructure or buildings that will have the least possible negative impact on the environment while still meeting the project's goals, a civil engineer and an ecologist can have a productive and collaborative relationship. Ecologists can offer insightful information about a site's natural elements, such as its water resources, wildlife habitats, and vegetation, which the civil engineer can use to guide their site investigation and to choose the project's best location on the site (or, if needed, select a different site).

Ecologists can assess the project's potential environmental impacts and pinpoint any ecological issues. The civil engineer can then create structures or infrastructure that takes these issues into account and reduces the project's total impact. Furthermore, the designing of green infrastructure can be done in collaboration with ecologists, especially with regard to designs of permeable pavement, rain gardens, and green roofs. Ecologists can also assist in supervising the construction process to make sure that environmental laws and best practices are followed, and that any effects on plants or wildlife are kept to a minimum.

Overall, cooperation between ecologists and civil engineers can aid in developing environmentally responsible and sustainable designs that strike a balance between societal requirements and the preservation of the natural world.

10.6 WORKING TOGETHER

Civil engineers are in charge of developing infrastructure, which is directly tied to many of the Sustainable Development Goals. For example, SDG 6 seeks to guarantee that everyone has access to clean water and sanitary facilities, and SDG 9 seeks to develop resilient infrastructure, advance sustainable industrialisation, and support innovation. SDG 11 strives to create inclusive, resilient, and sustainable cities and human settlements. Civil engineers must collaborate closely with other experts, such as town planners and environmental scientists, to accomplish these goals and, more broadly, to create integrated, sustainable infrastructure solutions that address community needs while reducing adverse environmental effects. Additionally, for infrastructure development to align with regional, national, and global sustainable development goals, civil engineers must also work with regulators and policymakers.

Besides being a necessity for achieving the SDGs, successful collaboration can also have several advantages. Civil engineers cannot finish a project by themselves, and so to guarantee that the project is developed and constructed to a high standard, they require the involvement and assistance of other professionals. Collaboration helps to ensure that everyone engaged is working towards the same objectives and assists in identifying and resolving any issues early on. Working together well with other experts can lead to unique and original answers to challenging engineering issues. Diverse viewpoints and areas of expertise can aid in identifying potential solutions and enhancing the project's overall quality. Collaboration with other professionals can also be important in identifying potential risks and creating mitigation strategies (Berglund et al. 2020).

Additionally, significant financial resources are often needed for civil engineering projects. Working together with other experts can be one way of finding affordable solutions, cutting down on waste (and therefore costs), and making the best use of the resources available. Other experts can also help with addressing regulatory requirements. Civil engineering projects must adhere to a variety of legal requirements, including safety standards, environmental laws, and building ordinances. Working with other experts can help to make sure the project complies with all applicable laws.

In conclusion, effective collaboration between experts is essential for civil engineers to control risk, maximise resources, and successfully complete projects whilst also adhering to legal and regulatory standards. For projects to be completed to the high standards of quality, safety, and sustainability, effective professional teamwork is necessary.

REFERENCES

Bauer, K. W. 2009. *City Planning for Civil Engineers, Environmental Engineers, and Surveyors*. Boca Raton: CRC Press.

Berglund, E. Z., J. G. Monroe, I. Ahmed, M. Noghabaei, J. Do, J. E. Pesantez, M. A. Khaksar Fasaee, E. Bardaka, K. Han, and G. T. Proestos. 2020. "Smart Infrastructure: A Vision For the Role of the Civil Engineering Profession in Smart Cities." *Journal of Infrastructure Systems* 26(2): 03120001.

Chan, E. H., M. Chan, D. Scott, and A. T. Chan. 2002. "Educating the 21st Century Construction Professionals." *Journal of Professional Issues in Engineering Education and Practice* 128(1): 44–51.

Christodoulou, S. 2004. "Educating Civil Engineering Professionals of Tomorrow." *Journal of Professional Issues in Engineering Education and Practice* 130(2): 90–94.

Hussin, A. A., and A. Omran. 2009. "Roles of Professionals in Construction Industry." The International Conference on Economics and Administration, Faculty of Administration and Business, University of Bucharest, Romania ICEA-FAA Bucharest.

Ozcep, F., and T. Ozcep. 2011. "Geophysical Analysis of the Soils For Civil (Geotechnical) Engineering and Urban Planning Purposes: Some Case Histories from Turkey." *International Journal of Physical Sciences* 6: 1169–1195.

Todoroff, E. C., T. Shealy, J. Milovanovic, A. Godwin, and F. Paige. 2021. "Comparing Design Thinking Traits between National Samples of Civil Engineering and Architecture Students." *Journal of Civil Engineering Education* 147(2): 04020018.

The Sustainable Future 11
Can We Still Make the Changes Today That We Should Have Made Yesterday?

11.1 INTRODUCTION

As an educator, I must believe that there is still time to implement the much-needed changes that will lead to a more sustainable future. If there is no hope, and the destruction of the planet is inevitable, then at best all we can seek to do is delay that outcome. I prefer to believe that working with the next generation of engineers will make a difference, that there are things we can do to lessen the harm and strive towards a more sustainable future.

Individual choices like cutting back on food waste, eliminating single-use plastics, conserving energy, and using more environmentally friendly transport modes can all make a difference. However, real structural changes that prioritise sustainability are needed, and must be backed up by regulations and laws. We must also teach ourselves and others about the importance of sustainability and how our actions affect the environment. We can develop a culture that values and prioritises sustainability for a better future.

DOI: 10.1201/9781003392194-11

With the help of the SDGs, which serve as a road map for achieving sustainability, we can ensure a more sustainable future for present and future generations. The SDGs emphasise the significance of group efforts and systemic transformation as necessary for achieving sustainability at the global scale. Civil engineers can help to create a more sustainable for everyone by developing and implementing sustainable infrastructure and supporting sustainable practices. Civil engineers can play a significant part in promoting sustainability through their work and their influence on society, even if they cannot save the future on their own.

11.2 TOOLS OF THE TRADE

There are a variety of tools and techniques available for civil engineers to use to ensure that projects are sustainable and responsive to their context. These tools range from assessments of how sustainable a project is through how to best engage with stakeholders.

For sustainability, the most common technique is a life cycle assessment (LCA), where civil engineers (or other professionals) assess the environment impact of a project at every stage of its life cycle, from the extraction of raw materials to manufacturing, transportation, use, and eventual disposal. An LCA can be used to find areas where changes can be made to lessen the effects on the environment; indeed, the goal in completing an LCA is to assess a project's embodied carbon, compare it to benchmark projects and industry standards, and offer suggestions for reducing the embodied carbon of the project. There are four main steps in the LCA process, beginning with the definition of the goals and scope, which are specific to the given project. Next, the life cycle inventory step involves gathering information on the inputs and outputs that are under evaluation, including the raw materials, energy, water, emissions, and waste. This step also considers the sourcing of materials and parts from suppliers and the embodied carbon in transporting the materials and parts to the construction site. Third, in the life cycle impact assessment step, the information that was gathered is used to assess the environmental effects of the project. Finally, in the interpretation step, recommendations for improvement are made, which can take into account the cost and availability of materials and parts, and the process may be repeated (see Figure 11.1). For detailed information on how to complete an LCA within a civil engineering context, see *Life Cycle Assessment* by Kathrina Simonen (Simonen 2014).

Architects, civil engineers, and construction professionals also use the digital tool known as Building Information Modelling (BIM) to design and manage infrastructure projects. With the use of BIM, designers and builders create a 3D virtual model of a structure to visualise, assess, and optimise its

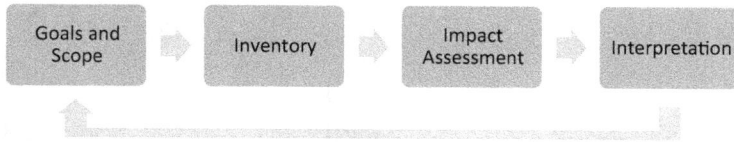

FIGURE 11.1 Life Cycle Assessment.

performance. The BIM model includes comprehensive details on the building's structural elements, such as the walls, floors, ceilings, doors, and windows, as well as details about its performance characteristics, such as thermal conductivity, acoustical qualities, and fire resistance.

Although BIM has been around since the 1970s, its application has substantially increased in recent years as a result of technological advancements and growing interest in sustainable building techniques (Chen et al. 2019). BIM was first primarily used for architectural designs but it has since developed into a comprehensive tool for managing a building or infrastructure project's whole lifecycle. Growing interest in sustainability has been one important driver of the trend towards using BIM in the construction industry (Charef 2022). BIM continues to become more popular and its use will likely continue to increase as the construction industry continues to adopt digital technology (Sacks and Barak 2010).

For a project to meet its stakeholders' needs, project managers and civil engineers need to understand how to prioritise the various stakeholders involved. A technique used to analyse and categorise stakeholders according to their influence and interest in a project is the Mendelow Matrix, sometimes referred to as Mendelow's Power-Interest Grid. It assists in identifying the most important stakeholders and in figuring out the best ways to interact and communicate with them. Based on the stakeholder's influence (or power) and interest, the matrix is divided into four quadrants, as shown in Figure 11.2. The High Influence, High Interest quadrant is for those stakeholders that hold the most power over the project and will need the most attention. They can utilise their influence to sway decisions and have a significant impact on the project. Regular interaction with these stakeholders and responsiveness to their concerns is key. The High Influence, Low Interest quadrant of stakeholders are those who could have significant influence over the project but their levels of interest are often low, so that they may need to be notified of changes but they do not require much attention or desire much involvement. The Low Influence, High Interest quadrant of stakeholders have little influence but much interest, and should be kept informed and involved even though they will not be able to derail a project. Lastly, the Low Influence, Low Interest stakeholders have little influence over or concern with the project, and they might need to be informed of the project but do not require much time or effort. By using the Mendelow Matrix, stakeholders can be prioritised and suitable communication and engagement strategies can be

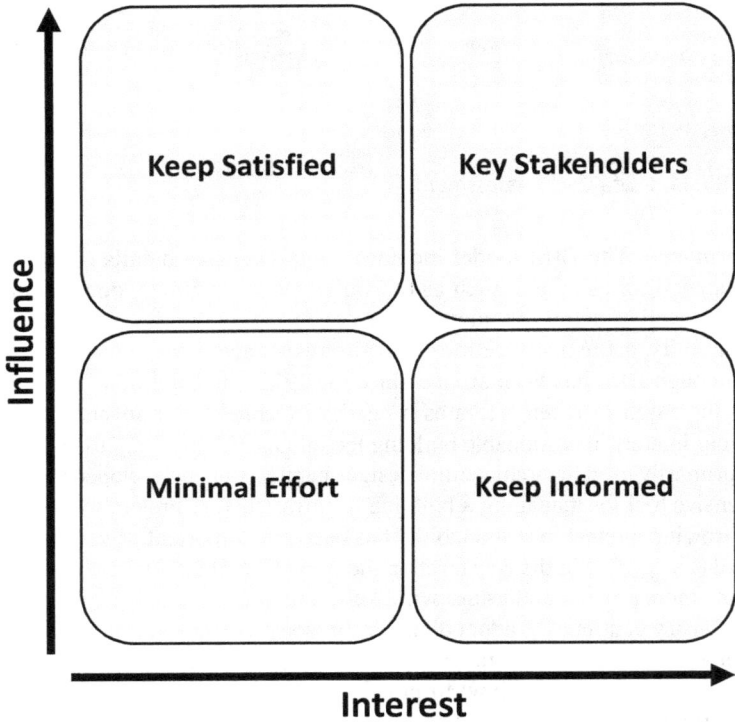

| Keep Satisfied | Key Stakeholders |
| Minimal Effort | Keep Informed |

Influence (vertical axis)

Interest (horizontal axis)

FIGURE 11.2   Quadrants of a Mendelow Matrix.

created for them. As projects evolve, stakeholders may move between quadrants and their levels of influence and interest may shift. Consequently, the matrix should be periodically examined and updated. For further explanation of how to use this technique, with examples set within a construction context, see Stefan Olander's article 'Stakeholder impact analysis in construction project management' (Olander 2007).

11.3 ESSENTIAL SKILLS FOR CIVIL ENGINEERS

Today's civil engineers are expected to have fluency in using various software packages, such as AutoCAD for creating plans, sections, and elevations of buildings and infrastructure, and GIS (geographic information systems) software to analyse and visualise information such as topography and land use. Additionally, project management software such as Microsoft Project can help civil engineers manage timetables, budgets, and resources. Although not a requirement, civil engineers within larger firms may use 3D printers to make models and prototypes of structures.

In addition to these technical skills, civil engineers need to have effective interpersonal skills, beginning with communication and their ability to explain technical concepts in a way that non-experts (e.g., clients, government officials, the general public) can comprehend. Civil engineers should also be able to analyse issues and come up with workable solutions – or what is commonly referred to as problem-solving skills. Furthermore, effective problem solving involves an element of creativity, where engineers need to be able to come up with original (sometimes unconventional) solutions to challenging issues.

Civil engineers frequently work in groups, and they should be able to work well with colleagues from a variety of specialties and backgrounds. They should be able to collaborate and to receive constructive criticism. They also need to be able to lead teams or projects, which in turn means being able to inspire and motivate their team members, successfully delegate tasks, and offer advice and assistance when required. Whether as the leader or as a member of a team, civil engineers need to be able to properly manage their time and prioritise their responsibilities, including when under the pressure of tight deadlines.

Finally, civil engineers must be able to adapt to new technologies, processes, contexts, and laws because the discipline of civil engineering is always changing. They should be willing to accept change and be open to evolving their skills.

11.4 INTERNATIONAL COLLABORATIONS

The future is likely to be increasingly international in nature. Large-scale infrastructure projects such as motorways, bridges, tunnels, airports, and seaports already may require civil engineers to travel abroad. These kinds of projects frequently involve more than one nation and call for international cooperation between engineers, builders, and governmental organisations. Environmental, geotechnical, and water resources engineering are additional civil engineering specialties that may require international collaboration.

Civil engineers may also have the opportunity to work on projects abroad if they work for international engineering firms, multinational corporations, or governmental organisations. Engineers who focus on areas like sustainability, risk analysis, or project management may also have more options to work abroad if they would like to do so.

Depending on the individual, their area of expertise, and the particular project that they are working on, the frequency with which civil engineers work internationally can vary greatly. Some civil engineers may routinely travel abroad whereas others may only sometimes or never work in an international context. Long-term career plans or temporary assignments may lead some civil engineers to work abroad for a period of time, and those who are interested in working in another country might look for opportunities through their professional networks, job advertisements, and international engineering organisations.

Although it is not a requirement for civil engineers to be multilingual, it can be useful in some circumstances, particularly if they work on international projects or with clients from other nations. Understanding local norms and laws, communicating with stakeholders more effectively, and creating better bonds with clients and partners from diverse backgrounds are all benefits associated with multilingualism. To be successful as a civil engineer, speaking multiple languages is not a must; many engineers work on domestic projects without ever having to speak a language other than their mother tongue. However, learning additional languages may be worthwhile, especially if an engineer aspires to work in international contexts.

11.5 THE WAY FORWARD

Civil engineers play a key role in future-proofing our cities and regions, and to play that role effectively they need to embrace sustainability, innovation, and adaptability. To prepare themselves as professionals, both as students and in their continuing professional development activities, they need to keep up with the newest technology and software, including tools for project management, simulations, and 3D modelling. With the speed at which technology is developing, civil engineering is becoming more and more dependent on fields like automation, machine learning, and artificial intelligence, and civil engineers will at least need to understand how developments in these adjacent fields can complement or enhance civil engineering. Staying up to date on the most recent advancements in civil engineering and adjacent fields is made easier by developing a strong network of connections and mentors, and keeping active in professional organisations. With a seemingly accelerating rate of change in technology, a mindset of adaptability and flexibility, being receptive to new concepts and methods, will help engineers to embrace change, learn from mistakes, and further their skills.

In their work, civil engineers must recognise that sustainability should be at the forefront of every project. Our cities are threatened by the climate crisis, which includes rising sea levels, severe weather events, and flooding. Building seawalls, enhancing stormwater management systems, and creating infrastructure that can survive extreme weather conditions are just a few examples of how civil engineers may future-proof our cities. Green infrastructure, such as bioswales and green roofs, can lessen the environmental impact of urbanisation through mitigating the urban heat island effect, boosting biodiversity, and improving air quality. Public transport, bike lanes, pedestrian-friendly roadways, and other sustainable transport options that alleviate traffic jams, enhance air quality, and encourage physical activity, can all be designed and implemented by civil engineers. Intelligent infrastructure solutions (e.g., real-time monitoring tools, data analytics, sensors) regulate traffic flow, cut down on energy use, and improve waste management – and again fall within the scope of civil engineering. Civil engineers will also continue to need to prioritise safety, which begins in the design process, continues through the maintenance of infrastructure, and includes the end-of-life demolition and disposal of projects.

Finally, communities can and should be actively involved in the design of infrastructure, and civil engineers should strive to understand the local context in which they are working, including the community's requirements, goals, and desires, through collaborating with neighbourhood organisations, holding meetings open to the public, and conducting surveys. Civil engineers can support the SDGs and help to prepare cities for the future by making sure infrastructure projects advance social, economic, and environmental sustainability and meet the needs of the community for this generation and for generations to come.

REFERENCES

Charef, R. 2022. "The Use of Building Information Modelling in the Circular Economy Context: Several Models and a New Dimension of BIM (8D)." *Cleaner Engineering and Technology* 7: 100414.

Chen, Y., Y. Yin, G. J. Browne, and D. Li. 2019. "Adoption of Building Information Modeling in Chinese Construction Industry: The Technology-organization-environment Framework." *Engineering, Construction and Architectural Management* 26(9): 1878–1898.

Olander, S. 2007. "Stakeholder Impact Analysis in Construction Project Management." *Construction Management and Economics* 25(3): 277–287.

Sacks, R., and R. Barak. 2010. "Teaching Building Information Modeling as an Integral Part of Freshman Year Civil Engineering Education." *Journal of Professional Issues in Engineering Education and Practice* 136(1): 30–38.

Simonen, K. 2014. *Life Cycle Assessment*. London: Routledge.

Index

For Product Safety Concerns and Information please contact our EU
representative GPSR@taylorandfrancis.com
Taylor & Francis Verlag GmbH, Kaufingerstraße 24, 80331 München, Germany

www.ingramcontent.com/pod-product-compliance
Lightning Source LLC
Chambersburg PA
CBHW061835220326
41599CB00027B/5292

9 7 8 1 0 3 2 4 9 1 1 4 1